Removal of Trace Contaminants from the Air

Victor R. Deitz, EDITOR
Naval Research Laboratory

A symposium co-sponsored
by the Division of Colloid
and Surface Chemistry and
the Division of Environmental
Chemistry at the 168th
Meeting of the American
Chemical Society, Atlantic
City, N.J., September
10-11, 1974.

ACS SYMPOSIUM SERIES **17**

AMERICAN CHEMICAL SOCIETY

WASHINGTON, D. C.   1975

Library of Congress ⌼P Data

Removal of trace contaminants from the air.

   (ACS symposium series; 17 ISSN 0097-6156)

   Includes bibliographical references and index.

   1. Air-Pollution—Congresses.   2. Air-Purification—
Congresses. 3. Atmospheric chemistry—Congresses.
   I. Deitz, Victor Revel, 1909-   II. American Chem-
ical Society. Division of Colloid and Surface Chemistry.
III. American Chemical Society. Division of Environ-
mental Chemistry. IV. Series: American Chemical Society.
ACS symposium series; 17.

TD881.R47   628.5'3   75-25568
ISBN 0-8412-0298-2   ACSMC8   17   1-207

ACS Symposium Series

Robert F. Gould, *Series Editor*

FOREWORD

The ACS Symposium Series was founded in 1974 to provide a medium for publishing symposia quickly in book form. The format of the Series parallels that of the continuing Advances in Chemistry Series except that in order to save time the papers are not typeset but are reproduced as they are submitted by the authors in camera-ready form. As a further means of saving time, the papers are not edited or reviewed except by the symposium chairman, who becomes editor of the book. Papers published in the ACS Symposium Series are original contributions not published elsewhere in whole or major part and include reports of research as well as reviews since symposia may embrace both types of presentation.

CONTENTS

Preface vii

1. Removal of Contaminants from Submarine Atmospheres 1
   Homer W. Carhart and Joseph K. Thompson

2. Heterogeneous Removal of Free Radicals by Aerosols in the
   Urban Troposphere 17
   L. A. Farrow, T. E. Graedel, and T. A. Weber

3. Precipitation Scavenging of Organic Contaminants 28
   R. N. Lee and J. M. Hales

4. Introduction, Transport, and Fate of Persistent Pesticides
   in the Atmosphere 42
   D. E. Glotfelty and J. H. Caro

5. Standards Development in the Control of Hazardous Contaminants
   in the Occupational Environment 63
   Douglas L. Smith and Jack E. McCracken

6. Electrically Augmented Filtration of Aerosols 68
   G. H. Fielding, H. F. Bogardus, R. C. Clark, and J. K. Thompson

7. Experimental and Theoretical Aspects of Cigarette Smoke Filtration 79
   Charles H. Keith

8. Aerosol Filtration by Fibrous Filter Mats 91
   William S. Magee, Jr., Leonard A. Jonas, and Wendell L. Anderson

9. Removal of Sulfur Dioxide in Stack Gases 106
   Donald A. Erdman

10. Kinetics of Trace Gas Adsorption from Contaminated Air 110
    Leonard A. Jonas, Joseph A. Rehrmann, and Jacqueline M. Eskow

11. The Reaction Between Ozone and Hydrogen Sulfide: Kinetics and
    Effect of Added Gases 122
    Sotirios Glavas and Sidney Tobey

12. Photolysis of Alkyl Nitrites and Benzyl Nitrite at Low
    Concentrations—An Infrared Study 132
    Bruce W. Gay, Jr., Richard C. Noonan, Philip L. Hanst, and
    Joseph J. Bufalini

13. **Fates and Levels of Ambient Halocarbons** 152
Daniel Lillian, Hanwant B. Singh, Alan Appleby, Leon Lobban,
Robert Arnts, Ralph Gompert, Robert Hague, John Toomey,
John Kazazis, Mark Antell, David Hansen, and Barry Scott

14. **The Fate of Nitrogen Oxides in Urban Atmospheres** 159
Chester W. Spicer, James L. Gemma, Darrell W. Joseph, and
Arthur Levy

15. **High Ozone Concentrations in Nonurban Atmospheres** 174
Lyman A. Ripperton, Joseph E. Sickles, W. Cary Eaton, Clifford
E. Decker, and Walter D. Bach

16. **Gas-Phase Reactions of Ozone and Olefin in the Presence of
Sulfur Dioxide** 187
D. N. McNelis, L. A. Ripperton, W. E. Wilson, P. L. Hanst, and
B. W. Gay, Jr.

Index 203

PREFACE

In the past the process flow of reactants to products has been open to varying extents, allowing some unused reactants, intermediates, and products to be released into the air. These materials may exert an objectional effect in themselves, or they may react under prevailing conditions to form undesirable products. Today, there is a strong effort to close the cycle by either imposing restrictions on particular processes or by interposing some chemical or physical control at the effluent interface.

The sixteen papers contained in this volume not only provide important and timely technical information but also focus attention on the great complexity of the problem of air pollution. One important aspect in current studies is the concern for the significant interactions among particulates and gas phase contaminants. The problems are difficult and numerous, and the solutions will require the application of the expertise in a diversity of disciplines. For this effort to be effective, the basic chemistry and physics of the air pollutants must be fully understood. The cooperation and assistance of the authors in the preparation of this volume has been outstanding and is gratefully acknowledged.

Naval Research Laboratory
Washington, D. C. 20375

Victor R. Deitz

Rutgers University
New Brunswick, N. J. 08903
July 11, 1975

Daniel Lillian

Removal of Contaminants from Submarine Atmospheres

HOMER W. CARHART and JOSEPH K. THOMPSON

U.S. Naval Research Laboratory, Washington, D. C. 20375

A submarine can be described very simply as a large bubble of air, encased in steel, and filled with machinery and men. This vessel is expected to go to sea and cruise under water for many weeks without access to surface air. The purpose of this paper is to describe some of the constituents of this large air bubble and then tell some of what is being done to maintain a livable atmosphere therein. A few problems that have arisen will be specifically mentioned.

The Chemistry Division of the Naval Research Laboratory has been engaged in submarine atmosphere research since 1929. With the advent of the nuclear submarine and its prolonged submergence capability about 1955, research on closed atmospheres has become an important field of study in the Navy (1, 2, 3, 4, 5, 6). Techniques developed in connection with nuclear submarine atmosphere research have also been extended and applied to atmosphere research for the Navy's SeaLab experiments. In the SeaLab experiments, men lived and worked for extended periods in and around a stationary underwater habitat at pressures determined by the depth of the water (7, 8). However, the work discussed here deals primarily with the nearly constant-pressure atmosphere of nuclear submarines. The progress that has been made results from the contributions of many people operating in a variety of scientific disciplines.

Composition of the Atmosphere in Underwater Craft

Table I, Column 1, shows the approximate composition of a typical nuclear submarine atmosphere. It is to be noted that, in addition to the normal constituents of air--nitrogen, oxygen, and a trace of argon, there are relatively small amounts of carbon dioxide, carbon monoxide, methane, hydrogen, and traces of a great variety of other compounds. These minor components cause most of the problems in maintaining air quality. Many of the

components are present in concentrations of a few parts per
million or less.

For comparison, Table I, Column 2, shows the approximate
composition of the atmosphere in SeaLab II, in which the total
pressure was typically 7 atmospheres. Man requires an oxygen
partial pressure of approximately 152 torr to sustain life,
and he can tolerate only certain limited partial pressures
of the various contaminants. In SeaLab II these partial pressure
limits are much lower percentages of the total pressure than
they are for the one atmosphere pressure of the nuclear submarine.

The techniques of gas and liquid chromatography and of
mass spectroscopy have been used in a continuing program of
identification and analysis of these trace contaminants in the
submarine atmosphere. Many of the contaminants, particularly
organic compounds of less than 10 carbon atoms, have been
quantitatively measured. Others have been qualitatively identified. Many more discrete components have been observed but not
identified. Over 400 discrete peaks were recorded in one gas
chromatogram of atmospheric contaminants collected on charcoal
and later extracted for analysis. The next several tables list
some of the compounds that have been found in nuclear submarine
atmospheres. These are grouped by chemical type for convenience
in listing and discussion. Among the hydrocarbons, aromatics
and their derivatives present the greatest problem from a
toxicological standpoint; hence, they have been studied more
thoroughly than some of the others (9, 10).

Table II shows a list of inorganic compounds that have been
identified in nuclear submarine atmospheres (9). Oxygen for
atmosphere replenishment is generated continuously at 3000
pounds per square inch by the electrolysis of water, thanks to
the ample supply of power available from the nuclear reactor.
For every mole of oxygen formed, two moles of hydrogen are
produced. This hydrogen must be pumped overboard. Any leakage,
of course, goes into the ship's atmosphere. Another source of
hydrogen, and also of arsine and stibine, is the bank of lead-
acid storage batteries which the ship carries as an emergency
power source.

Carbon dioxide is a product of respiration. Most of this is
absorbed from the air by a monoethanolamine (MEA) scrubber. Spent
amine is subsequently heated to release the CO_2, which is pumped
overboard. The regenerated amine is then cooled and recycled
through the scrubber. In time, this MEA partially breaks down
to form ammonia. Both ammonia and MEA vapor in the air are
oxidized in the catalytic burner (see below) to form oxides of
nitrogen. Hydrogen fluoride is a product of decomposition of
certain refrigerant gases in the catalytic burner. Ozone and
nitrogen dioxide are produced by improperly functioning
electrostatic precipitators, present to remove aerosols from
the air. Carbon monoxide is a major product of smoking, as are
aerosols and a variety of other pyrolysis products from tobacco.

The catalytic burner is a flow-through reactor whose bed is packed with Hopcalite, a catalyst consisting of coprecipitated oxides of copper and manganese. This catalyst, when heated to 315°C, quantitatively oxidizes H_2, CO, and many organic compounds to CO_2 and H_2O. Methane is an exception; it is not oxidized appreciably at ordinary operating temperatures. The catalytic burner's effect on certain other compounds will be discussed later in this paper.

Table III lists aliphatic hydrocarbons that have been identified in nuclear submarine atmospheres (9). Except for methane, these come largely from paints, sealers, and cements that continue to exude vapors long after application and from solvents, oils, and fuels. Methane is a major contaminant in the submarine atmosphere. The principal sources of methane are flatus and the decomposition of wastes in the ship's septic tank. Methane is not appreciably oxidized in the catalytic burner; hence, its concentration continues to increase as long as the ship is closed. In a prolonged closed period the methane concentration may reach 500 parts per million or more. Fortunately, this concentration is well below the 5.3% minimum concentration required for flammability in air. Also, methane is relatively non-toxic; so its presence is of no great concern.

Table IV shows a list of aromatic hydrocarbons that have been quantitatively measured in nuclear submarine atmospheres (9, 10). These also can be traced to solvents, oils, cements, and sealers, as well as to by-products of smoking, cooking, and decomposition of hot oils in ship's machinery. Aromatics make up about 25% of the total hydrocarbon content of the submarine atmosphere.

Table V lists some of the unsaturated and alicyclic compounds that have been identified in submarine atmospheres (9).

Table VI shows other miscellaneous compounds that have been found in submarine atmospheres (9). Monoethanolamine, of course, comes from the CO_2 scrubber. Ethyl alcohol comes largely from the medical department. Certain others can be traced to paint and cement solvents.

Table VII shows some of the chlorinated compounds that have been identified (9). These have been particularly troublesome-- not only because of their own toxicity, but also because some are decomposed or converted in the catalytic burner to form other harmful materials. A particular example, methyl chloroform, will be discussed later in this paper.

Table VIII lists refrigerant and propellant gases that have been found on board submarines. These gases come from leaks in the refrigeration systems and from pressurized aerosol containers that have been brought on board (9).

The foregoing tables show a large number of compounds that have been identified in submarine atmospheres. These are not isolated cases. Figure 1 shows a comparison of the concentrations

Table I. Atmospheres in undersea craft;
approximate values, dry basis

Gas	Nuclear Submarines	SEALAB II
N_2	78%	<18%
O_2	19-21%	3.5-5.0%
CO_2	0.8-1.3%	<0.4%
Ar	0.9%	0.1%
He	0	>78%
CO	<25 ppm (ave.)	30 ppm (meas.max.)
CH_4	0-600 ppm	44 ppm (max.)
Higher organics	30 mg/m^3 (ave.)	15 mg/m^3 (STP) (max.)
H_2	0-0.35%	0

Table II. Inorganics identified in submarine atmospheres

Oxygen	Chlorine
Nitrogen	Hydrogen fluoride
Hydrogen	Nitrogen dioxide
Carbon Dioxide	Nitrous oxide
Carbon Monoxide	Ozone
Water	Stibine
Ammonia	Sulfur dioxide
Arsine	

Table III. Aliphatic hydrocarbons identified
in submarine atmospheres

Methane	n-Heptane
Ethane	n-Octane
n-Butane	n-Nonane
i-Butane	n-Decane
n-Pentane	n-Undecane
i-Pentane	n-Dodecane
n-Hexane	

Table IV. Aromatic hydrocarbons identified
           in submarine atmospheres

Benzene	1,3,5-Trimethylbenzene
Toluene	1,2,4-Trimethylbenzene
o-Xylene	1,2,3-Trimethylbenzene
m-Xylene	Indane
p-Xylene	n-Butylbenzene
Ethylbenzene	i-Butylbenzene
n-Propylbenzene	sec-Butylbenzene
i-Propylbenzene	tert-Butylbenzene
o-Ethyltoluene	Diethylbenzene
m-Ethyltoluene	1,3-Dimethyl-5-ethylbenzene
p-Ethyltoluene	C_5-Alkylbenzenes
	Naphthalene

Table V. Unsaturates and alicyclics identified
          in submarine atmospheres

Acetylene	i-Butene
Ethylene	Isoprene
Propylene	n-Decene
1-Butene	n-Undecene
2-Butene (cis)	Methylcyclohexane
2-Butene (trans)	Ethylcyclohexane

Table VI. Miscellaneous organics identified
           in submarine atmospheres

Acetaldehyde	Methyl alcohol
Acetic acid	Methyl ethyl ketone
Acetone	Methyl isobutyl ketone
Ethyl acetate	Monoethanolamine
Ethyl alcohol	Phenol
Formaldehyde	i-Propyl alcohol

of ten aromatic hydrocarbons found in six different submarines.
The general similarity between atmospheres from different ships
is obvious, although each ship tends to have its own "signature"
(10).
        Figure 2 draws another comparison. Here the aromatic
hydrocarbon content of the submarine atmosphere (average of six
ships) is compared to the aromatic hydrocarbon yield from
petroleum distillate (average of seven samples). The implica-
tion is that diesel fuel and paint thinners (both of which are
petroleum distillates) may account for a major share of these
hydrocarbons in the submarine atmospheres (10).
        A major source of pollution in the submarine atmosphere is
smoking (9, 11). In addition to the smoke, carbon monoxide
and pyrolysis products must also be considered. Table IX shows
some of the products of burning tobacco (12). The main stream
from a cigarette is that which the smoker inhales and exhales
himself. The side stream is that which rises from the cigarette
completely independent of the smoker. All these products are
evolved into the atmosphere of the boat, and any person present
becomes a smoker whether he wants to or not. Electrostatic
precipitators remove some of the smoke particles but the gases
and vapors remain in the general atmospheric mix. Figure 3
shows the aerosol concentration versus time during one submarine
cruise (11). The decrease in concentration at 60 hours due to
a temporary smoking ban was quite dramatic, as was the later
recovery when the ban was lifted and all the smokers caught
up. There appeared to have been some cheating during the ban,
as evidenced by the gradual rise in aerosol concentration
between 80 and 120 hours.

Contamination Control in Submarines

        Many of the contaminants described in the foregoing
section are in low concentration and would cause little concern
in an ordinary work-day situation on land, where one could open
a window or turn on a ventilator fan. However, in the sealed
environment of a submarine, where exposure goes on for 24 hours
a day for many weeks, the allowable limits must be kept very low.
        One of the first places to attack a potential contamination
problem is at its source by restricting the use of possible
hazardous materials either in construction of a ship or in its
later operation. It is for this reason that the Navy has adopted
the use of water-thinned paints for interior painting where
possible in place of petroleum-based paints. In addition,
painting is restricted to periods when the ship can be open
during the drying process. Aerosol propellants are a great
convenience for dispensing toiletries such as shaving cream
or deodorant. But if 120 men each carried on board two 4-ounce
cans of aerosol product containing 50% propellant, there would
be 30 pounds of propellant released into the ship's atmosphere.

Table VII.   Chlorinated compounds identified
in submarine atmospheres

Chloroform	Vinyl chloride
Methyl chloride	Vinylidene chloride
Dichloromethane	Dichloroethylene
Carbon tetrachloride	Trichloroethylene
Methyl chloroform	Tetrachloroethylene

Table VIII.   Refrigerants and aerosol propellants
identified in submarine atmospheres

Code	Formula
R-11	CCl_3F
R-12	CCl_2F_2
R-113	CCl_2FCClF_2
R-114	$CClF_2CClF_2$
R-114B2	$CBrF_2CBrF_2$
–	$CHCl_2CF_3$
–	CHF_2CH_2Cl

Table IX.   Products from regular size cigarette (mg/cig)
as determined with a Standard Smoking Machine

Product	Main Stream	Side Stream
H_2O	7.9	250
CO_2	55.2	524
Organics	10.3	69.8
CO	17.2	39.4

(Total CO from King size cigarette ca. 80 mg/cig)

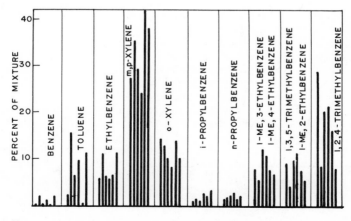

Figure 1.   Comparison of aromatic hydrocarbons from atmospheres of six submarines

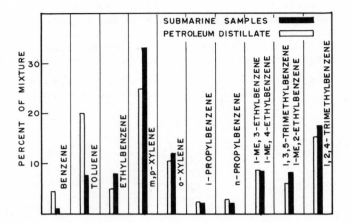

Figure 2. Relative distribution of aromatic hydrocarbons recovered from submarine atmospheres (average of six samples) compared to petroleum distillates (average of seven)

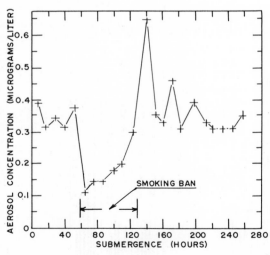

Figure 3. Aerosol concentration in a submarine atmosphere during a cruise. Note effect of temporary ban on smoking.

So the use of these products is restricted so far as possible. The Navy maintains a list of allowable materials which may be brought on board. Anything not on that list is contraband. Even so, there may be some smuggling.

The four contamination-removal systems on board a nuclear submarine are (1) the MEA scrubber for CO_2 removal, (2) electrostatic precipitators for removal of aerosols, (3) the catalytic burner for oxidation of H_2, CO, and organic compounds, and (4) an adsorbent carbon bed for removal of large-molecule organic compounds. These systems are essentially independent of each other. Each processes only a portion of the ship's air at a single pass; eventually, all the air is processed, and equilibrium is reached as mixing goes on.

The MEA scrubber maintains CO_2 levels between 0.8 and 1.2%. This is higher than the normal open atmospheric CO_2 concentration (0.03%), but apparently it can be tolerated. The possibility of subtle physiological effects of CO_2 at these concentrations must be considered, however. At present, there is no other regenerative CO_2 removal system that has been developed to a degree suitable for use in submarines. However, improved and new techniques are being studied extensively (1, 4).

Electrostatic precipitators serve well for aerosol removal when they are properly maintained. However, precipitators are sometimes installed in locations that are virtually inaccessible for cleaning and servicing. As a result, when such devices lose efficiency or fail, they are often merely shut down for the duration of a cruise (3).

The catalytic burner is quite effective for converting H_2, CO, most hydrocarbons (except methane), and oxygenated organic compounds to CO_2 and H_2O. The burner is less effective at its usual operating conditions for oxidizing nitrogen- and chlorine-containing compounds. Table X shows the relative efficiency of conversion of some typical nitrogen compounds (13, 14). These are partially oxidized to produce N_2O. A certain amount of NO_2 is also produced. A concurrent problem arose in the early days with the then available gas analyzer. N_2O interfered with the instrumentation for measuring CO concentration, giving a spurious high value for CO. However, this was not all bad, for the crew's normal response to a high CO value was to run the burner longer, thus more thoroughly oxidizing the contaminants.

Halogen-containing compounds may be converted by the Hopcalite catalyst to other compounds which are more of a hazard than the original (14, 15, 16). Table XI shows the percentage decomposition of some halogenated compounds in the catalytic burner. Some refrigerants are partially converted to halogen acids. For this reason, an alkaline adsorbent, Li_2CO_3, is used to clean up the effluent from the catalytic burner. Other refrigerants are decomposed only slightly if at all. The concentration of these stable compounds increases with time as long as the ship is closed.

   A particularly troublesome problem was caused by methyl chloroform in the catalytic burner. Methyl chloroform was a non-flammable solvent used in the cement that was used to secure the insulation to the hull surfaces of submarines. This cement continued to evolve solvent vapors for a long time after application. The methyl chloroform was being converted in the catalytic burner to vinylidene chloride and trichlorethylene as shown:

$$
\begin{array}{l}
\overset{\overset{\text{H}\quad\text{Cl}}{\vdots\quad\vdots}}{\underset{\text{H}\quad\text{Cl}}{\text{H}-\text{C}-\text{C}}}-\text{Cl} \longrightarrow
\end{array}
\left[
\begin{array}{l}
\underset{\text{H}\quad\text{Cl}}{\text{H}-\text{C}=\text{C}}-\text{Cl} \quad\text{(70\% vinylidene chloride)}\\[2em]
\underset{\text{Cl}\quad\text{Cl}}{\text{H}-\text{C}=\text{C}}-\text{Cl} \quad\text{(5\% trichlorethylene)}\\[2em]
\text{traces of others.}
\end{array}
\right.
$$

Both these products are much less desirable than their precursor. Vinylidene chloride is not only more toxic, but it is also more resistant to further oxidation than its precursor. Here is a case of the submarine's air purification equipment manufacturing additional contamination.

   The fourth contamination removal system, the activated carbon adsorbent bed, adsorbs large-molecule organic compounds (6, 17). Thus, it helps to remove odors from the air and serves as a sink for hydrocarbons and other organic compounds which are normally condensable at room temperature. In so doing, the carbon bed also prevents overtaxing the catalytic burner. The carbon bed consists of about 500 pounds of activated charcoal packed in 5-pound cloth bags. These small bags are stacked sand-bag fashion in a plenum in the air duct to form a filter bed. Such a construction may allow a considerable amount of leakage; one such bed when tested showed 50% leakage. Leakage of the carbon bed is not too serious a problem, however, because the bed processes only a portion of the total air supply at a single pass.

   In a submarine it is impractical to attempt to monitor the atmosphere for all known contaminants. The Naval Research Laboratory has developed a specialized gas chromatograph, called the NRL Total Hydrocarbon Analyzer (18). This chromatograph uses a reversible column with ship's air as the carrier gas. First, with the column operating in a forward mode, the instrument gives a resolved analysis for methane, refrigerants, and "light ends" that are not normally adsorbed by activated carbon. Then the flow is reversed, and the column is back-flushed. The ensuing single peak represents the concentration of all other organic vapors. The use of air as a carrier gas

avoids adding still another component to the atmospheric burden.
This Total Hydrocarbon Analyzer is carried on many nuclear
submarines and is used for analysis of the organic vapor content
of the atmosphere. Figure 4 shows a typical chromatogram from
this instrument.

Table XII shows the overall performance of the organic
vapor removal system (catalytic burner plus carbon bed) during
a series of tests (19). It is to be seen that vapors were
generated from various sources in the ship at a nearly constant
rate of about 700 grams per day. The overall rate of removal
was also approximately constant, although the load distribution
shifted as the various components were operated. The equilibrium
vapor concentration decreased as additional burner/adsorbent
capacity was provided. The catalytic burners accounted for
considerably more organic vapor removal than did the carbon bed.

As a carbon bed continues to adsorb vapors, it eventually
reaches equilibrium with the ambient concentration in air and
is no longer effective as an adsorbent. At present, the carbon
beds on submarines are replaced on an arbitrary schedule without
knowledge as to their remaining adsorptive capacity. One of the
Laboratory's recent tasks has been to develop a method to assess
the condition of the carbon bed in situ and thus be able to know
when the bed should be replaced.

A method is being developed which uses the NRL Total Hydrocarbon Analyzer (20). At regular intervals, organic vapor
concentrations will be determined in both the effluent and the
influent air streams of the carbon bed. The ratio of effluent
to influent vapor concentration should remain relatively
constant as long as the carbon bed is functioning properly.
When the carbon begins to lose its effectiveness, this effluent
to influent ratio should show an upward trend. This upward
trend would be the signal to replace the carbon bed.

Bench-scale studies have been made in which samples of
carbon were preloaded with various amounts of n-decane, a heavy,
low-volatility hydrocarbon. These samples were then challenged
with low concentrations of n-hexane in air to represent the
more volatile components of a submarine atmosphere. It was
found that initial break-through of n-hexane occurred at about
7% total hydrocarbon loading, regardless of the relative
amounts of the two test hydrocarbons. Results of these bench-
scale studies are summarized in Figure 5. The method is now
under test aboard ship to determine whether or not it is practical
for routine use.

New Areas For Research

An area of interest related to atmosphere control in
submarines is that of fire fighting. A fire aboard any ship
is serious, but in a closed ship such as a submarine, a fire
can easily become a disaster. Not only the combustion products,

Table X. Hopcalite-catalyzed oxidation of nitrogen compounds

Compound	Inlet Concentration (ppm)	N_2O Formed (% theoretical)
Ammonia	140	70
MEA	19	20
Morpholine	35	16
Pyridine	40	17

NOTES: (1) Space velocity 21,000 hours^{-1} and temperature
            315°C.
        (2) Raising the furnace temperature diminished the
            yield of N_2O from MEA and Ammonia.

Table XI. Decomposition of halogenated compounds over
           Hopcalite at 315°C.

Code	Formula	Decomposition (%)
R-11	CCl_3F	30-50
R-12	CCl_2F_2	<1
R-14	CF_4	0
R-22	$CHClF_2$	25
R-114B2	$CBrF_2CBrF_2$	30
R-C318	C_4F_8 (cyclic)	0
Methyl chloroform	CH_3CCl_3	>98

Table XII. Generation and removal rates of organic
            contaminants in USS Sculpin

			"Total Hydrocarbon" Content		
Phase	Date	Removal Agent	Equil. conc, mg/m^3	Removal Rate, grams/day	Generation Rate, grams/day
I	2-21-63	Burner 1	100	700	700
II	2-27-63	Burner 2	60	420	600
		Carbon Bed		180	
III	3-2-63	Burner 1	40	280	680
		Burner 2		280	
		Carbon Bed		120	

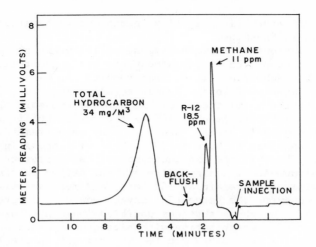

Figure 4. Chromatogram of organic vapor content in submarine atmosphere as determined by NRL Total Hydrocarbon Analyzer

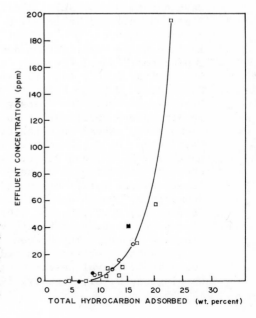

Figure 5. Effluent n-hexane concentration as a function of total hydrocarbon adsorbed. □—weighted values for n-decane and n-hexane. ○—weighted value for n-decane; value for n-hexane calculated from concentration-time product. ■, ●—values for carbon bed with built-in leak.

but also most known fire fighting agents can be hazardous
in a closed environment.

The relationship between oxygen partial pressure and
percentage composition of the atmosphere has suggested a possible
way to combat submarine fires. Man requires an oxygen partial
pressure of about 152 torr. In our normal atmosphere, this is
about 20% of the total pressure. At a total pressure of 1.5
atmospheres, 152 torr is only 14% of the total, but it is
sufficient for man's needs. However, many open flame fires
which will continue to burn in 20% oxygen will be extinguished
at 12-14%. Thus, if one can increase the total pressure in a
closed environment by adding nitrogen, the oxygen partial
pressure will still remain at 152 torr, and man can still
survive. However, the percent of oxygen would be lowered
to values low enough so that combustibles would not burn.
This concept of pressurization with nitrogen to snuff out fires
in submarines is now being studied extensively in the laboratory
and is showing a high degree of promise (21, 22, 23).

Literature Cited

1. Miller, R. R., and Piatt, V. R., Eds., "The Present Status
   of Chemical Research in Atmosphere Purification and
   Control on Nuclear-Powered Submarines," Naval Research
   Laboratory Report 5465, Washington, D. C., April 1960.
2. Piatt, V. R., and Ramskill, E. A., Eds., "Annual Progress
   Report: The Present Status of Chemical Research in
   Atmosphere Purification and Control on Nuclear-Powered
   Submarines," Naval Research Laboratory Report 5630,
   Washington, D. C., July, 1961.
3. Piatt, V. R., and White, J. C., Eds., "Second Annual
   Progress Report: The Present Status of Chemical Research
   in Atmosphere Purification and Control on Nuclear-Powered
   Submarines," Naval Research Laboratory Report 5814,
   Washington, D. C., August, 1962.
4. Carhart, H. W., and Piatt, V. R., Eds., "Third Annual
   Progress Report: The Present Status of Chemical Research
   in Atmosphere Purification and Control on Nuclear-Powered
   Submarines," Naval Research Laboratory Report 6053,
   Washington, D. C., December, 1963.
5. Lockhart, L. B., and Piatt, V. R., Eds., "Fourth Annual
   Progress Report: The Present Status of Chemical Research
   in Atmosphere Purification and Control on Nuclear-Powered
   Submarines," Naval Research Laboratory Report 6251,
   Washington, D. C., March, 1965.
6. Alexander, A. L., and Piatt, V. R., Eds., "Fifth Annual
   Progress Report: The Present Status of Chemical Research
   in Atmosphere Purification and Control on Nuclear-Powered
   Submarines," Naval Research Laboratory Report 6491,
   Washington, D. C., January, 1967.

7. Umstead, M. E., and Smith, W. D., "Trace organic contaminants
    in the atmosphere of SEALAB II; preliminary report on,"
    Naval Research Laboratory letter report 6180-89:MEU:WDS:ec,
    26 May 1966.
8. Umstead, M. E., "Carbon monoxide in the atmosphere of
    SEALAB II; information on," Naval Research Laboratory letter
    report 6180-55A:MEU:ec, 31 March 1966.
9. Carhart, H. W., and Piatt, V. R., Eds., "Third Annual Progress
    Report: The Present Status of Chemical Research in Atmosphere
    Purification and Control on Nuclear-Powered Submarines,"
    Naval Research Laboratory Report 6053, Chapter 8, Washington,
    D. C., December, 1963.
10. Johnson, J. E., Chiantella, A. J., Smith, W. D., and
    Umstead, M. E., "Nuclear Submarine Atmospheres: Part 3 –
    Aromatic Hydrocarbon Content," Naval Research Laboratory
    Report 6131, August, 1964.
11. Piatt, V. R., and Ramskill, E. A., Eds., "Annual Progress
    Report: The Present Status of Chemical Research in
    Atmosphere Purification and Control on Nuclear-Powered
    Submarines," Naval Research Laboratory Report 5630, Chapter
    18, Washington, D. C., July, 1961.
12. Bitner, J. L., Naval Research Laboratory, Washington, D. C.,
    Personal communication.
13. Carhart, H. W., and Piatt, V. R., Eds., "Third Annual
    Progress Report: The Present Status of Chemical Research in
    Atmosphere Purification and Control on Nuclear-Powered
    Submarines," Naval Research Laboratory report 6053, Chapter
    9, Washington, D. C., December, 1963.
14. Christian, J. G., and Johnson, J. E., "Catalytic Combustion
    of Nuclear Submarine Atmospheric Contaaminants," Naval
    Research Laboratory Report 6040, Washington, D. C., March
    1964.
15. Musick, J. K., and Williams, F. W., Ind. Eng. Chem., Prod.
    Res. Develop., (1974), 13, 175-179.
16. Musick, J. K., and Williams, F. W., "Catalytic Decomposition
    of Halogenated Hydrocarbons with Hopcalite Catalyst,"
    American Society of Mechanical Engineers Publication
    75-ENAs-17, New York, N. Y., 1975. Presented at Intersociety
    Conference on Environemntal Systems, San Francisco,
    California, July 21-24, 1975.
17. Johnson, J. E., "Nuclear Submarine Atmospheres: Analysis
    and Removal of Organic Contaminants," Naval Research
    Laboratory Report 5800, Washington, D. C., September 1962.
18. Eaton, H. G., Umstead, M. E., and Smith, W. D., J.
    Chromatographic Sci., (1973), 11, 275-278.
19. Umstead, M. E., Smith, W. D., and Johnson, J. E., "Submarine
    Atmosphere Studies Aboard USS Sculpin," Naval Research
    Laboratory Report 6074, Washington, D. C., February, 1964.
20. Eaton, H. G., Thompson, J. K., and Carhart, H. W.,
    "Feasibility of the Total Hydrocarbon Analyzer for Evaluating

the Life of Charcoal Beds," Naval Research Laboratory Report
7712, Washington, D. C., April, 1974.

21. Carhart, H. W., and Gann, R. G., "Fire Suppression in
Submarines," Report of NRL Progress, May, 1974.

22. Tatem, P. A., Gann, R. G., and Carhart, H. W., Combustion
Sci. and Technol., (1973), 7, 213.

23. Tatem, P. A., Gann, R. G., and Carhart, H. W., Combustion
Sci. and Technol., in press.

Heterogeneous Removal of Free Radicals by Aerosols in the Urban Troposphere

L. A. FARROW, T. E. GRAEDEL, and T. A. WEBER

Bell Laboratories, 600 Mountain Ave., Murray Hill, N. J. 07974

The effect of aerosols on atmospheric photochemistry has been evaluated in a computation of the gas phase chemistry of the urban troposphere for the northern New Jersey-New York metropolitan region. It is shown that the heterogeneous incorporation of free radicals into atmospheric aerosols provides an efficient radical sink, and also stabilizes the diurnal variation of each radical so that it is reproducible from day to day. For the inorganic free radicals $N\dot{O}_3$ and $H\dot{O}$, which play an important role in oxidizing NO to NO_2, it is shown that a large overnight storage occurs when aerosols are omitted from the calculation. Because of the importance of radical reaction chains in the atmospheric chemistry of stable gases, the diurnal behavior of the stable species is also influenced by aerosol interactions. Thus, the extensive data on stable gases may be used to study the effect of the presence of the aerosol. The computational results for NO and NO_2 obtained with the inclusion of aerosols are shown to be in better agreement with the field data than are the results obtained with their omission. It is concluded that this heterogeneous chemical component is an essential part of the chemistry of the troposphere.

I. INTRODUCTION

While it is generally accepted that free radicals play a very important role in the gas phase chemistry taking place in contaminated urban atmospheres, it has not been well recognized thus far that the concentrations and behavior of these free radicals are very much influenced by the presence of aerosol particles in the atmosphere. It will be shown in this paper that these aerosols act as an important heterogeneous removal mechanism for free radicals, and that this removal affects the concentrations and diurnal patterns in the urban troposphere of such stable gases as NO and NO_2.

These conclusions have been reached as a result of an extensive computation which has given special emphasis to understanding the details of the gas phase chemistry of the contaminated atmosphere as represented by a set of 127 chemical reactions in 73 species. The basis for selection of this set and the accompanying rate constants has been extensively documented elsewhere [1].

Further, these computations have been performed with specific reference to the conditions prevailing in the northern New Jersey-New York metropolitan region. The computation was performed for three counties in northern New Jersey, those being, from west to east, Morris, Essex and Hudson. Bulk air transport between counties was included in accordance with prevailing measured wind patterns, and within each county according to the daily variation in mixing height. Emission rates of primary contaminants from motor vehicle traffic, industrial activity, and other sources, were made available from inventories on a county wide basis for the three counties mentioned above. Finally, the diurnal variation of photon flux from the sun for the particular latitude and time of the year was considered; in this work only sunny summer days were included. Complete specification of all of these boundary conditions has been presented elsewhere [2].

II. AEROSOLS IN URBAN ATMOSPHERES

The tropospheric aerosol is an ensemble of particles of different shapes, sizes, and chemical compositions. Its ubiquity, together with its large surface-to-volume ratio, suggests that heterogeneous reactions between gas-phase species and aerosols must be included in a comprehensive chemical description of the atmosphere [3-9]. The specification of the interaction between aerosols and gas-phase species has historically proved difficult, principally because the estimation of bulk activation energies for adsorption of gas-phase molecules and radicals on particle surfaces of unknown composition is currently intractable [10-13]

The key to inclusion of heterogeneous atmospheric chemistry is the realization that pure inert solid particles may be a rarity in the atmosphere, and that the chemically inert particle core is in most or all cases surrounded by a hygroscopic layer [14]. This layer, upon contact with water vapor, will form a solution with an equilibrium suitably depressed from that of pure water such that the aqueous layer is stable. Analyses commonly show that water comprises 20-70 percent of the weight of the atmospheric aerosol and that the aerosol size spectrum is a function of relative humidity [15-22]. Further, aerosols typically possess electric charge [23-25], and the tendency of water to cluster readily about charged particles is well established [26,27]. This information indicates that atmospheric aerosols may be accurately represented as solid core particles with water shells rather than as particles with solid surfaces. Thus, the aerosol-gas interaction may be treated by gas-liquid, as opposed to gas-solid, interface techniques. The solubilities of the atmospheric gases in the water-shelled aerosol can be estimated by assuming each dissolved gas to be in equilibrium with its vapor, and applying Henry's law to find the total dissolved volume. This gives upper limits for the quantities of the stable species that can be absorbed in the liquid layer. The results indicate that the amounts of atmospheric gases absorbed in the aerosol are negligible fractions of the total gas-phase concentrations [1] and can thus be neglected for atmospheric chemical purposes.

The interactions between radicals and water-covered aerosols are likely to be of substantially more interest than is the case with the stable gases. It appears certain that the inorganic radicals such as HO_2 and HO will be instantly absorbed on impact; this is also extremely likely for other radicals. In our "first order" formulation, therefore, it is assumed that *all* radicals will be absorbed into aerosols on impact, a conclusion that has been reached independently by Warneck [28]. The resulting concept for aerosol/gas interaction is thus that indicated on Figure 1. (There is experimental evidence for the rapid destruction of HO on solid surfaces [29], in which case the formulation we propose may be independent of the surface properties, rather than requiring a liquid surface.)

The frequency of collisions of gas radicals with a surface is given by

$$f = A \left[\frac{kT}{2\pi M} \right]^{\frac{1}{2}} [X] \qquad (1)$$

where A is the total surface area, $[X]$ is the radical concentration, M is the free radical mass, and the remaining terms within the parentheses have their standard gas kinetic meanings [30]. The removal rate for radicals may thus be expressed (at $T = 25\,^{\circ}C$) by

$$R = \frac{0.085 \, P[X]}{\sqrt{M}} \text{ ppm min}^{-1} \qquad (2)$$

where the integrated cross-sectional area for an atmospheric aerosol with a typical size spectral distribution [20] and a known concentration [31] has been utilized for the area, and where P represents the actual aerosol concentration (in $\mu g/m^3$) in the reaction volume of interest. Extensive observations of aerosol size spectra in Pasadena [20] have been utilized to define the diurnal variation in aerosol concentrations; these data are consistent with somewhat less comprehensive studies at Whippany, New Jersey (T. E. Graedel, unpublished data, 1973). Figure 2 shows the smooth analytic function that was fit to the data and used in this computation.

III. MAGNITUDE AND STABILIZATION EFFECTS OF AEROSOL INCLUSION

The expression for the removal rate developed above was included in the differential equation for each of the free radicals partaking in the chemistry of the contaminated troposphere. However, in order to assess the effect of this formulation, the calculation was rerun omitting the aerosol removal mechanism completely. In the former case, the expected result of lower magnitude of free radicals was verified, but in addition it was found that the aerosol interaction has the effect of stabilizing the diurnal pattern of free radicals such that it repeats over several days. Since this stability is a requirement for a credible formulation, the aerosol inclusion is thus of vital importance.

The two effects described above are graphically illustrated in Figure 3, which shows the diurnal variation, over a period of simulated real time of 48 hours, for NO_3 and HO_2, two radicals which are important in the oxidation of NO to NO_2. The curves of Figure 3a, which were calculated with the inclusion of aerosol effects, show good repeatability, whereas the curves of Figure 3b show great differences from one day to the next. It is also clear that the magnitudes shown in Figure 3a are less than those shown in Figure 3b, as expected.

The reason for the stabilizing effect of the aerosols may be understood by looking at the total rate of formation of NO_3 from chemical reactions alone and at the rate of removal of NO_3 by aerosols. The difference between these two rates gives the total rate, which determines the overall behavior of the NO_3 concentration. These three rates are displayed in Figure 4. (The advection effect due to bulk air flow from Essex to Hudson county is negligible here since the magnitudes of NO_3 are comparable in both counties and thus as much of this species is advected in from

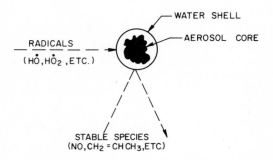

Figure 1. Schematic of the water-shell aerosol
model for impact removal of radicals

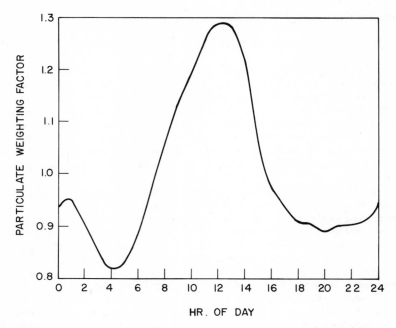

Figure 2. The diurnal variation in the aerosol concentration weighting
factor

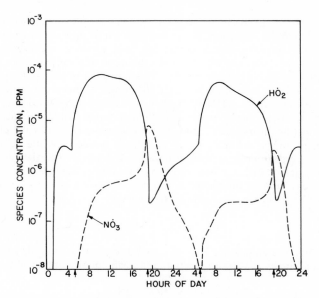

Figure 3a. Computed two-day diurnal concentration patterns for the HȮ$_2$ and NȮ$_3$ radicals in Hudson County, N. J. Arrows on the abscissa of this and following figures indicate times of sunrise and sunset.

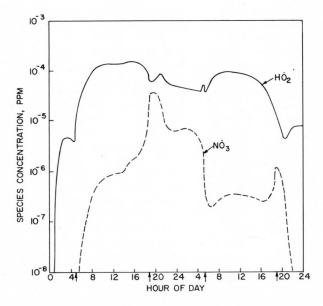

Figure 3b. Same as Figure 3a for calculations in which the heterogeneous removal of radicals by the atmospheric aerosol was not included

Essex as is advected out of Hudson towards Manhattan. In some cases, however, advection effects may be important, as has been shown in the case of ozone [32].) It is immediately obvious from Figure 4 that the $N\dot{O}_3$ removal rate nearly follows the chemical production rate, with the difference being so small that it must be displayed on a different scale in order to be visible. This mirroring of the chemical rate by the aerosol rate is a direct consequence of Eq. (2), which shows that the rate is directly proportional to the amount of $N\dot{O}_3$ present, with some variation due to the diurnal change in the aerosol concentration itself. Similar curves can be produced for all the free radicals in this formulation. Thus any day to day variation in the chemical production rate, as shown in Figure 4, will tend to be balanced by the aerosol removal. In particular, the daytime peak concentrations are depleted overnight to give an approximate return to initial conditions existing 24 hours earlier.

The late day peak in the chemical production rate is easily explained when it is realized that $N\dot{O}_3$ is produced chiefly by the reaction

$$NO_2 + O_3 \rightarrow N\dot{O}_3 + O_2 \qquad (2\text{-}7)^*$$

As the sun sets, the amount of NO_2 rises because photodissociation into NO and O diminishes and finally ceases. In addition, the main chemical removal rate

$$N\dot{O}_3 + NO \rightarrow 2NO_2 \qquad (2\text{-}11)$$

drops late in the day as the level of NO falls with the decrease of NO_2 photodissociation. Thus the very sharp rise is produced by the simultaneous increase in $N\dot{O}_3$ production and decrease in removal. The rise is short-lived since the ozone necessary for production in (2-7) falls off rapidly as the sun sets because the atomic oxygen necessary for ozone production is no longer produced by photodissociation of NO_2.

Figure 5 presents the curve for the total chemical rate of formation of $N\dot{O}_3$ when aerosols are not considered. It should be carefully noted here that the total chemical rate is negative rather than positive and that the negative peak occurs three hours earlier than the positive peak in the computation with aerosols (Figure 4). This difference is caused by the change in the diurnal behavior of NO_2 with and without aerosols.

In the latter case, NO_2 drops faster and reaches an afternoon low value which is a factor of five less than that obtained with aerosols (Figure 6). This is sufficient to allow the removal rate of (2-11) to dominate the formation rate of (2-7) to give the total negative value.

Whereas the advection term played a negligible role in the aerosol case of Figure 4, it is obvious from Figure 5 that this term is important in the absence of aerosols. Since the advection rate is proportional to the difference in the concentrations of $N\dot{O}_3$ in Essex and Hudson counties, the positive rate indicates more $N\dot{O}_3$ in Essex than in Hudson when aerosols are neglected, whereas the magnitudes are about equal with aerosols and follow approximately the same diurnal pattern. Without aerosols, the diurnal patterns in Essex and Hudson are not stabilized and correlate poorly. The larger $N\dot{O}_3$ magnitude in Essex is caused by the presence of smaller concentrations of NO for removal and larger concentrations of O_3 for production

---

*Numbering of reactions follows that of Reference 1.

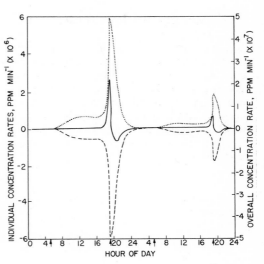

Figure 4.   Comparative rates of NȮ₃ concentration variations resulting from gas phase chemistry (. . . . .), heterogeneous combination with aerosols (– – –), and the total of all source and sink rates (————). The first two individual rates are referred to the labeling of the left-hand ordinate, the total rate to the labeling of the right-hand ordinate.

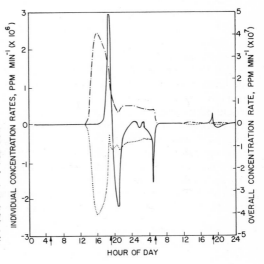

Figure 5.   Comparative rates of NȮ₃ concentration variations resulting gas phase chemistry (. . . . .), advection (– · – · –), and the total of all source and sink rates (————) for calculations in which the heterogeneous removal of radicals by the atmospheric aerosol was not included. The first two individual rates are referred to the labeling of the left-hand ordinate, the total rate to the labeling of the right-hand ordinate.

than exist in Hudson. The aerosol removal term helps to keep the magnitudes more constant from county to county, but without this removal the inequality exists and the advection rate term becomes comparable to the chemical rate term.

IV. COMPARISON WITH FIELD DATA

Within a county, Figure 3b shows a strong overnight "storage" in the atmosphere of $N\dot{O}_3$ and $H\dot{O}_2$ without the removal mechanism of the aerosols, whereas the presence of aerosols steadily depletes these two free radicals to a low morning value, thus assisting in the day to day reproducible pattern of Figure 3a. The behavior of these two free radicals, while not measured directly, may be assessed indirectly by observing their effects upon NO and NO_2 and then comparing with the extensive field data which is available for these two stable gases.

Our results indicate that the free radical oxidation equations in order of importance are

$$NO + R\dot{O}_2 \rightarrow NO_2 + R\dot{O} \quad (3\text{-}4),(3\text{-}13),(3\text{-}16),(4\text{-}5),$$
$$(4\text{-}9),(4\text{-}12),(6\text{-}8),(7\text{-}6),(8\text{-}8)$$

$$NO + H\dot{O}_2 \rightarrow NO_2 + H\dot{O} \quad (2\text{-}4)$$

$$NO + N\dot{O}_3 \rightarrow 2NO_2 \quad (2\text{-}11)$$

where $R\dot{O}_2$ is a generic indication of oxidized hydrocarbon free radicals. Thus the connection between the two inorganic free radicals of Figure 3 and NO and NO_2 is an intimate one, particularly since the sum of the rates of the previous 3 equations is comparable in magnitude to the difference between the principle inorganic balance

$$NO_2 \overset{h\nu}{\rightarrow} NO + O \quad (2\text{-}5)$$

$$NO + O_3 \rightarrow NO_2 + O_2. \quad (2\text{-}2)$$

Figure 6 shows the results for NO and NO_2. In each case, the calculated curve with (solid) and without (dashed) aerosols is shown, along with the data which have been stratified according to the conditions of time of year (summer), prevailing wind direction, wind speed, day of the week, and integrated solar radiation. NO shows only a single peak, which results from the morning rush hour injection. Since NO is a primary product of combustion, its behavior is determined less by chemistry than by sources. Even here, however, the greater magnitude of the peak is more in keeping with the median values of the data. NO_2, on the other hand, is dominated by chemical production and destruction, and has a good deal of structure. The early morning fall in the data is adhered to by the case with aerosols only, and the mid-afternoon trough agrees much better for the case with aerosols. Atmospheric data for ozone has also been shown to support the aerosol interaction calculation [33].

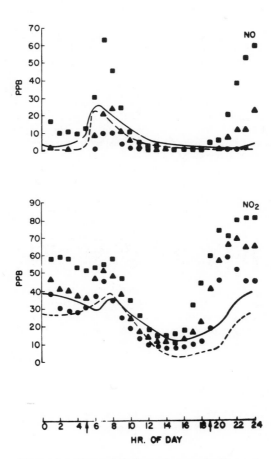

Figure 6. Computed diurnal concentration patterns [calculation with aerosols (————), without aerosols (— — —)] for Hudson County, N. J., compared with measured concentration distribution parameters (see text): □, upper quartile; △, median; ○, lower quartile. The observational data was measured at Bayonne, N. J. by the New Jersey Department of Environmental Protection.

V. CONCLUSIONS

A computation of the gas phase kinetics of the urban troposphere has been made which includes the effects of inhomogeneous reactions with atmospheric aerosol particles. The aerosol particle is adduced from calculational and experimental evidence to have an inert core surrounded by a water shell. As a first approximation stable gases may be considered in equilibrium with this water shell, a situation that results in negligible effects on gas-phase concentrations. Reasons are advanced to indicate that free radicals colliding with aerosols will be absorbed upon impact. The calculated effect of this removal mechanism on free radicals is to decrease the radical concentrations at a rate proportional to their absolute magnitudes, and to deplete overnight the daytime peak concentrations such that the initial conditions established twenty four hours earlier are closely reproduced. For the important inorganic free radicals NO_3 and HO_2, it is shown that omission of aerosols from the calculation causes a large overnight storage of these two species. Direct experimental field data on free radicals are not available for comparison. However, the level and behavior of the free radicals in turn affect the behavior of stable gases such as NO and NO_2 for which ample field data are available, and it is with this indirect verification that the validity of the aerosol inclusion is tested and found effective.

LITERATURE CITED

1. Graedel, T. E. and Farrow, L. A., submitted for publication, 1975.
2 .Weber, T. A., Graedel, T. E. and Farrow, L. A., submitted for publication, 1974.
3. Junge, C., *Air Chemsitry and Radioactivity,* (New York: Academic Press), 1963.
4. Cooke, L. M., ed., *Cleaning Our Environment: the Chemical Basis for Action,* American Chemical Society, Washington, D. C., p. 32ff, 1969.
5. Urone, P. and Schaeffer, G. A., *Environ. Sci. Technol.,* **3,** 681, 1969.
6. Bufalini, M., *Environ. Sci. Technol,* **5,** 685, 1971.
7. Cadle, R. D., *J. Colloid Interface Sci.,* **39,** 25, 1972.
8. Altshuller, A. P., *Environ. Sci. Technol.,* **8,** 709, 1973.
9. Calvert, J. G., paper presented at the Conference on Health Effects of Air Pollutants, National Academy of Sciences, Washington, D. C. October 3-5, 1973.
10. Pilat, M. J., *J. Air Poll. Cont. Assoc.,* **18,** 751, 1968.
11. Fraser, D. A., *Adsorption of Sulfur Dioxide on Particulate Surfaces,* API Publication 4079, American Petrolum Institute, Washington, D. C., 1971.
12. Judeikis, H. S. and Siegel, S., *Atmos. Environ.,* **7,** 619, 1973
13. Siegel, S. and Judeikis, H. S., paper presented at Midwest Meeting, Amer. Chem. Soc., Chicago, Illinois, August, 1973.
14. Winkler, P., *J. Aerosol Sci.,* **4,** 373, 1973.
15. Orr, C., Hurd, F. K., Hendrix, W. P. and Junge, C., *J. Met.,* **15,** 240, 1958.
16. Ludwig, F. L. and Robinson, E., *Tellus,* **22,** 94, 1970.
17. Abel, N., Winkler, P. and Junge, C., *Studies of Size Distributions and Growth with Humidity of Natural Aerosol Particles,* Max-Planck-Institut fur Chemie, Mainz, Germany (NTIS Document AD 689 189), 1969.
18. Lundgren, D. A. and Cooper, D. W., *J. Air Poll. Cont. Assoc.,* **19,** 243, 1969.
19. Meszaros, A., *Tellus,* **23,** 436, 1971.

20. Husar, R. B., Whitby, K. T. and Liu, B. Y. H., *J. Colloid Interface Sci.*, **39**, 211, 1972.
21. Rosinski, J. and Nagamoto, C. T., *J. Colloid Interface Sci.*, **40**, 116, 1972.
22. Whitby, K. T., Husar, R. B. and Liu, B. Y. H., *J. Colloid Interface Sci.*, **39**, 177, 1972.
23. Gillespie, T., *Electric Charge Effects in Aerosol Particle Collision Phenomena, Aerodynamic Capture of Particles,* (E. G.. Richardson, ed), (New York: Pergamon Press), p.44, 1960.
24. Whitby, K. T. and Liu, B. Y. H., *The Electrical Behavior of Aerosols, Aerosol Science,* (C. N. Davies, ed.), (New York: Academic Press), p. 59, 1966.
25. Takahashi, T., *J. Atmos. Sci.,* **29**, 921, 1972.
26. McKnight, L. G., *Trans. Am. Geophys. Union,* **53**, 391, 1972.
27. Kiang, C. S., Stauffer, D. and Mohnen, V. A., *Nature,* **244**, 53, 1973.
28. Warneck, P., *Tellus,* **26**, 39, 1974.
29. Mulcahy, M. F. R. and Yound, B. C., paper presented at *Symposium on Chemical Kinetics Data for the Lower and Upper Atmsophere,* Warrenton, Va., September 18, 1974.
30. Moelwyn-Hughes, E. A., *Physical Chemistry,* (New York: Pergamon Press), p. 45, 1961.
31. Lundgren, D. A. *J. Colloid Interface Sci.,* **39**, 205, 1972.
32. Graedel, T. E., Farrow, L. A. and Weber, T. A., *Proc. Symp. on Atmos. Diffusion and Air Pollution,* (American Meteorological Society, Boston, Mass.), p. 115-120, 1974.
33. Graedel, T. E., Farrow, L. A., and Weber, T. A., *Int. J. Chem. Kinetics,* (in press), 1975.

3

Precipitation Scavenging of Organic Contaminants

R. N. LEE and J. M. HALES

Battelle, Pacific Northwest Laboratories, Richland, Wash. 99352

To the casual urban observer the effect of precipitation on visibility and airborne dust is readily apparent. Atmospheric cleansing via this important process is thus widely appreciated. Quantitative study of precipitation scavenging is historically associated with the development of nuclear energy and attendent concern for the fate of airborne radionuclides. These studies continue to be of interest both from the standpoint of a growing nuclear industry and a more general awareness of man's impact on the atmospheric environment. This awareness has lead to quantitative descriptions of polluted atmospheres, demand for a greater understanding of pollutant transport and finally to the complex question of possible weather modification effects. In spite of a general concern for a host of atmospheric contaminants, the area of gas scavenging continues to be virtually ignored. Prior to the recent extensive examination of SO_2 washout it was generally assumed that precipitation scavenging of gases could be understood and predicted using concepts derived from aerosol scavenging theory. Although gas solubility was felt to be of theoretical importance in determining washout efficiency, the premise of infinite solubility appeared to be nearly universally accepted in discussing the washout behavior of trace gases and vapors.([1]) The limited applicability of this assumption was dramatically illustrated by the results derived from the study of precipitation interaction with an elevated SO_2 plume.([2]) Sulfur dioxide concentrations well below those anticipated and the frequent lack of a correlation between plume location and the distribution of SO_2 among collected rain samples were among the observations recorded during this study. These findings forced a closer examination of the interaction of rain with this gaseous contaminant, including consideration of contaminant desorption from the falling raindrop. The study of SO_2 washout confirmed the importance of this phenomenon and culminated in the development of the EPAEC gas washout model.

The work presented in this paper is an extension of the
aforementioned investigation and was designed to verify application of the EPAEC model to trace organic vapors under general
precipitation conditions. The tracers employed in this study were
the moderately soluble organics ethylacetoacetate and
diethylamine.

Fundamental Features of the EPAEC Model

The fundamental equation of the EPAEC scavenging model is

$$\frac{dc}{dz} = \frac{3K_y}{V_t a} (y_{Ab} - H'c) \quad , \tag{1}$$

which describes the concentration change, $\frac{dc}{dz}$, in a raindrop of
radius   a   falling through a gaseous concentration field denoted
by $y_{Ab}(x,y,z)$. In equation 1, H' is an appropriate solubility
parameter and K is a mass-transfer coefficient, which may be
estimated using standard techniques. V_t denotes the terminal fall
velocity of the raindrop. Reversibility of the system is provided
by the term $(y_{Ab}-H'c)$ which may be either positive (absorption) or
negative (desorption) depending upon the relative concentrations
of material in the drop and the local gas phase.
    The EPAEC code provides numerical approximations of solu-
tions to (1) for a number of drop sizes at a given ground-level
receptor location. The resulting values are then weighted according to the appropriate raindrop size spectrum, and utilized
to compute an average concentration of material in the rain. A
complete description of this code is available elsewhere.[3]

Experimental Design

    A. Test Site. The field tests described in this paper were
conducted at the Quillayute airfield on the Olympic Peninsula of
western Washington. This site offered the advantages of ample
rainfall, a U.S. Weather Bureau station and isolation from urban
air pollution sources. While much of this region is heavily
wooded, the airfield provided a sizeable area free of structures
which might introduce irregularities in the air motion over the
sampling network. In addition, the stand-by status of the airfield permitted uninterrupted use of this facility.
    The distribution of rain samplers with respect to the source
is shown in Figure 1. Arc A, at 300 feet, consisted of 17 sampling sites located at 5° intervals within the 80° arc defined by
lines drawn NNE(20°) and WNW(300°) from the tower. Arc B consisted of 29 sample locations encompassing the same angle as arc
A and following the geometry of the airfield runways. Although
winds during each test carried the center of the plume through
the region described by arcs A and B, additional samplers were set

out during experiments 2 and 3 to insure enclosure of the plume.

    B. Meteorological Sensors and Data Handling. Meteorological data pertinent to application and appraisal of the EPAEC model included wind velocity and direction, rainfall rate and raindrop size distribution. A Gill anemometer was located at the top of the release tower to measure the required wind data. The u, v, and w wind components were recorded at 0.42 second intervals on Metrodata tape. This data was later transferred to industry compatible 7-track tape using a NOVA computer before submission to a UNIVAC computer for calculation of the mean wind velocity, direction and standard deviation in the vertical and azmuthal wind directions. In addition to translating the original tape, this intermediate step incorporated a check on data quality which permitted rejection of faulty data.

    Raindrop spectra were obtained during the test period by brief exposure of water sensitive paper (ozalid paper). Those images taken during a period of representative rainfall (as suggested by rainfall rate data) were examined using a Zeiss spectrometer. The sizes of raindrop images were related to actual drop size by a previous calibration and the measured population distribution was used to characterize the test period.

    C. Tracer Release Equipment. Release of the organic plume was from a vapor dispersion system constructed for this program. It consisted of a 5-gallon storage tank, a compressed air cylinder and a vapor gun constructed by mounting two sonic nozzles inside an 8-inch aluminum tube. Evaporation of tracer emerging from the sonic nozzles was aided by warming the air mass receiving it using a heat gun mounted directly behind the nozzles. During operation of the system, the organic liquid was transferred under low pressure to the sonic nozzles and injected into the atmosphere by high pressure air from the air compressor. Figure 2 shows operation of the release system during one of the field experiments. In this instance, as in all other field tests, the source was located adjacent to the Gill anemometer at the top of a 100 foot tower.

    D. Rain and Vapor Samplers. Rain samples were collected in polyethylene bottles. The complete sampling system consisted of a 1-liter bottle fitted with a 7-3/8 inch polyethylene funnel. The bottle was surrounded with dry ice and thermally insulated by enclosure in a polystyrene container. This design permitted the rain to be frozen on impact reducing the potential for tracer loss via desorption to the atmosphere and adsorption to container walls. Laboratory tests conducted prior to the field experiments indicated that dilute aqueous solutions of both tracers could be stored at room temperature for a period of at least one month without suffering a significant concentration change. Sample storage at sub-freezing temperatures was adopted however as an

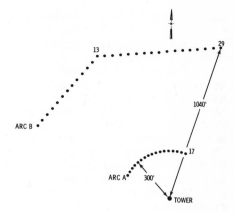

Figure 1.   *Configuration of test grid*

Figure 2.   *Operation of vapor dispersion
equipment during field experiment*

added precaution. In addition to the collection of rain samples,
the integrated vapor phase concentration of tracer at ground
level was determined through the use of a limited number of vapor
samplers. These units consisted of a battery-operated pump which
drew air at 1 lpm through an impinger containing 10 ml of distilled water (ethylacetoacetate release) or 10 ml of $5x10^{-4}$ N nitric
acid (diethylamine release). At the conclusion of an experiment,
the impinger solutions were transferred to polyethylene bottles
for storage. Analysis of the impinger solutions gave the total
mass of tracer in the air sampled and provided an indication of
the relative importance of dry deposition and washout to the
tracer load in the collected rain samples. Since only a limited
number of the vapor samplers were available, their distribution
among the sampling sites was dictated by wind conditions prior to
tracer release. Further mention will not be made of these
samples since subsequent analysis verified the minimal contribution of dry deposition to the contaminant load of the collected
rain samples.

    E. Tracer Analysis. Analysès of ethylacetoacetate and
diethylamine were accomplished by standard colorimetric methods.
Both procedures were sufficiently sensitive to detect concentrations of less than 0.1 ppm and were easily performed in the field.
This feature was of particular significance in that it allowed
immediate analysis of selected samples and hence appraisal and
modification of the field program if required. Although on-site
analyses of rain and impinger samples were carried out, the bulk
of the analytical work was completed after transport of the
frozen samples to the Battelle Hanford Laboratory. Rain and
impinger samples were readied for analysis by bringing them to
room temperature before the removal of an aliquot. Analytical
precision was determined to be better than 30% by multiple
analysis of standards and selected field samples.
    Ethylacetoacetate was determined by formation of the tran-
sient enol-diazonium addition compound.(4) This species shows
an absorption maximum at 435 nm which varies linearly with concentration to approximately 1 ppm.
    Diethylamine was converted to the chloramine and subsequently
allowed to react with starch-iodine solution to yield the intense
blue color of the starch-tri-iodide complex.(5) While not
specific for diethylamine, background rain and vapor samples confirmed the absence of interference by local amine sources. This
procedure yields a linear Beer's Law plot to approximately 0.8ppm.

    F. Tracer Solubility. Since accurate solubility data has
not been documented for the vapor concentrations of interest,
estimates were employed to predict the rainwater concentration of
tracer along the sampling arcs. The estimated solubilities, based
on measurements made at high vapor concentration and available
vapor pressure data are expressed in terms of Henry's Law

$$H' = \frac{\text{contaminant vapor pressure}}{\text{aqueous concentration of contaminant}} \qquad (2)$$

Although the solubility of diethylamine is certainly a function of pH, this dependence was ignored in the present application of the EPAEC Model and the Henry's Law constant for the amine assumed to be a true constant. The value of H' for diethylamine was calculated to be 11 atm cc/mole using the high concentration data of Dailey and Felsing,(6) the vapor pressure data of Perry(7) and the Clausius-Clapeyron equation. Ethylacetoacetate solubility was calculated with less certainty by taking the vapor pressure data of Perry and the handbook values of water solubility to yield an H' value of 0.25 atm cc/mole.

Field Experiments

In this section we will provide a brief description of the meteorological conditions influencing each of the scavenging experiments. Concentration data will also be presented for each test and later compared with calculated tracer concentrations derived from use of the EPAEC model.

a. Experiment 1. This test was divided into two segments when rain ceased midway through the test period. Both segments were characterized by a variable rainfall rate which averaged 0.20 cm/hr. Winds throughout the test were out of the south-southeast at 9 mph with occasional gusts of up to 20 mph. A total release of 9.8 liters of ethylacetoacetate during the 18-minute test period gave a source strength of 3.36 gm/sec. Tracer concentrations in the collected rain samples are displayed graphically in Figure 3. The maximum concentrations found in arcs A and B were 1.6 (A-8) and 1.2 (B-13) ppm respectively.

b. Experiment 2. This test, conducted on the same day as test 1, was 24 minutes in length. Winds continued to be moderate (7 mph average with gusts up to 22 mph) shifting somewhat to the southeast from those experienced earlier in the day. In response to the more easterly flow of surface winds, arcs A and B were extended by 15° with the addition of three rain samplers. Rainfall during the period was uniform averaging 0.59 cm/hr. Ethylacetoacetate was again employed as the tracer. Dispersal of 11.4 liter of this material gave a source strength of 4.6 gm/sec. Figure 4 summarizes tracer concentration data for this test. Maximum concentrations observed in arcs A and B were 2.0 (A-6) and 0.37 (B-12) ppm respectively.

c. Experiment 3. This test was the first involving the use of diethylamine. The higher volatility of this material permitted an increase in the source strength which in this test was

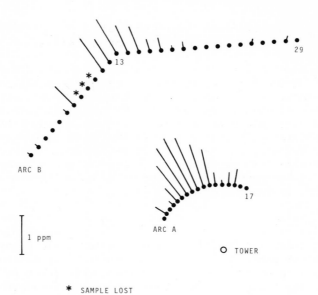

Figure 3. Measured concentrations of ethylacetoacetate in rainwater, run 1

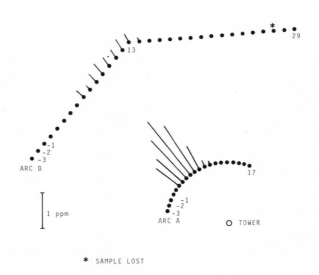

Figure 4. Measured concentrations of ethylacetoacetate in rainwater, run 2

22.1 gm/sec. The 40-minute experiment was characterized by an average rainfall rate of 0.25 cm/hr although significant variation was observed throughout the period. Winds were moderate averaging 8 mph from the south-southeast and gusting to as high as 26 mph. Surface winds were again observed to be light and variable causing modification of the sampling grid to accommodate possible washout by rain falling to the east of the sampling grid. The distribution of rainwater concentrations are illustrated graphically in Figure 5. Maximim concentrations in arcs A and B were 2.5 ppm (A-10) and 3.0 ppm (B-15) respectively.

d. Experiment 4. This 30-minute test was characterized by a nearly uniform rainfall rate of 0.25 cm/hr. Strong winds at 16 mph moved the plume slightly to the west of north. Although the standard deviation in horizontal wind direction was less than observed for any other test, a relatively broad distribution of tracer was found in samples collected in arc B. This is apparently a result of the influence of high winds on the y component of the raindrop trajectory. Graphical presentation of the relative concentrations of diethylamine in the collected rain samples is shown in Figure 6. Peak concentrations occurred at sites A-10 (1.3 ppm) and B-17 (2.2 ppm).

e. Experiment 5. Test 5 was conducted under the same general meteorological conditions as encountered during test 4. Winds at source level were southerly at 16 mps while surface winds appeared to be from the south-southwest. Rainfall throughout the 25-minute experiment was heavy decreasing significantly only during the final 4 minutes. The average precipitation rate was 1.1 cm/hr falling primarily as rain but mixed with sleet or snow during the initial 3-4 minutes of the test. Rainwater concentrations are summarized in Figure 7. The large volumes of rain collected and higher winds encountered during this test are reflected in lower maximum concentrations of 0.57 ppm (A-12) and 0.63 ppm (P-18) for arcs A and B respectively.

Model Predictions

The rainwater concentrations recorded graphically in the preceeding 5 figures are of interest in comparison to those calculated by application of the EPAEC model. Several assumptions have been made in assigning values to many of the parameters required for these calculations and an analysis of the possible error associated with each has identified the solubility term, H', as introducing the greatest uncertainty. The solubilities of ethylacetoacetate and diethylamine are currently under study and will permit a rigorous evaluation of the EPAEC model when they become available. Although an evaluation of the model using current test results is not yet possible, it is of interest to compare the observed concentrations with those anticipated for various

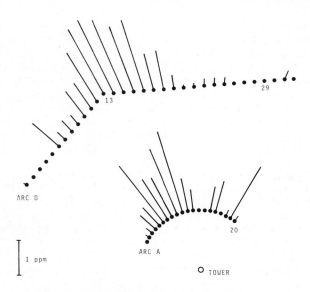

*Figure 5.   Measured concentrations of diethylamine
in rainwater, run 3*

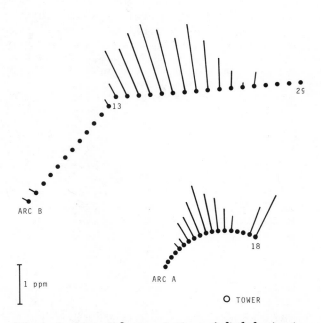

*Figure 6.   Measured concentrations of diethylamine in
rainwater, run 4*

tracer solubilities.

The data summarized in Table I pertains to experiment 1. It is immediately apparent that an assumption of infinite tracer solubility (H'=0) is inappropriate for ethylacetoacetate. Observed tracer concentrations are also substancially less than predicted using the estimated H' value of 0.25 atm cc/mole. Examination of similar data for experiment 2, Table II, also suggests that ethylacetoacetate may be less soluble than predicted by the rough calculations outlined earlier. In each of these tests reasonably good agreement between observed and predicted concentrations is realized if an H' value of 4 atm cc/mole is used to define solubility.

As with the ethylacetoacetate tests, the assumption of infinite diethylamine solubility yields predicted peak concentrations which are generally an order of magnitude greater than observed. Initial results derived from the investigation of diethylamine solubility indicate a Henry's Law constant for this material of approximately 6 atm cc/mole. Assignment of this value to the solubility parameter yields predicted concentrations between those recorded for infinite (H'=0) and low (H'=11) solubility. Comparisons for the diethylamine experiments are found in Tables III - V. Although predictions based on finite tracer solubility show fair agreement with the observed concentrations in the case of test 3, a similar comparison for experiments 4 and 5 is less satisfying. Poorer agreement is, however, to be expected in view of the more severe meteorological conditions accompanying these tests. Both were characterized by strong variable winds. These conditions have been observed during previous SO_2 studies to result in poorer agreement between observed and calculated concentrations, particularly for sampling sites located close to the source. This is interpreted to be a consequence of plume undercutting by the rain and deficiencies in the plume model.

Conclusions

This paper has focused on the washout behavior of organic vapors which may be characterized as being moderately soluble in water. Although we recognize the absence of accurate solubility data essential to the complete evaluation of this work, the following tentative conclusions appear to be justified.

1. The occurrence of reversible contaminant-rainwater interaction, first recognized in the study of SO_2 washout, is surely important to the removal of ethylacetoacetate and diethylamine. Contaminants possessing similar physical properties may therefore be assumed to exhibit like behavior.

2. The ability of the EPAEC model to successfully predict real world behavior is encouraging.

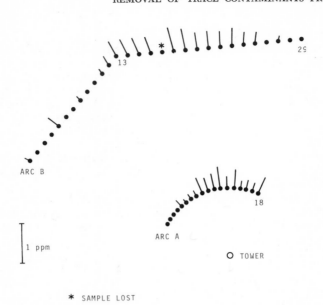

*   SAMPLE LOST

Figure 7.   Measured concentrations of diethylamine in rainwater, run 5

Table I.   Observed Tracer Concentration vs.
Concentrations Predicted for Various Solubilities —
Experiment 1 (EAA)

STATION	C_{obs}	$C_{H' = 0}$	$C_{H' = 0.25}$	$C_{H' = 4}$
A 3	< 0.10 ppm	0.04 ppm	0.04 ppm	0.02 ppm
4	0.16	0.19	0.18	0.10
5	0.50	0.35	0.71	0.34
6	1.1	2.61	2.41	0.94
7	1.5	9.10	7.96	1.96
8	1.6	36.2	27.7	2.97
9	1.5	16.6	13.9	2.63
10	1.2	4.98	4.48	1.44
11	1.0	0.79	1.12	0.52
12	0.32	0.26	0.25	0.14
13	0.10	0.05	0.05	0.02
B 8	----	0.06	0.05	0.02
9	----	0.24	0.21	0.07
10	----	0.80	0.67	0.20
11	1.0	2.11	1.68	0.43
12	0.70	4.32	3.18	0.71
13	1.2	6.39	4.43	0.91
14	0.78	5.90	4.21	0.87
15	0.73	3.29	2.54	0.58
16	0.33	1.48	0.97	0.27
17	0.35	0.31	0.27	0.09
18	0.12	0.07	0.06	0.02

Table II. Observed Tracer Concentration vs. Concentrations Predicted for Various Solubilities — Experiment 2 (EAA)

STATION	C_{obs}	$C_{H' = 0}$	$C_{H' = 0.25}$	$C_{H' = 4}$
A 1	< 0.10 ppm	0.29 ppm	0.27 ppm	0.14 ppm
2	< 0.10	0.94	0.89	0.43
3	0.80	2.6	2.38	1.03
4	1.0	5.8	5.23	1.97
5	1.5	12.0	9.99	2.94
6	2.0	50.3	28.7	3.39
7	1.6	12.0	9.99	2.94
8	0.77	5.8	5.23	1.97
9	0.20	2.6	2.38	1.03
10	0.11	0.94	0.89	0.43
11	< 0.10	0.29	0.27	0.14
12	< 0.10	0.07		
13	< 0.10	0.02		
B 5	< 0.10	0.01		
6	0.26	0.05		
7	0.16	0.19	0.16	0.06
8	0.31	0.62	0.49	0.17
9	0.36	1.7	1.23	0.38
10	0.31	3.7	2.44	0.68
11	0.27	6.2	3.73	0.97
12	0.37	7.5	4.31	1.09
13	0.28	6.2	3.73	0.97
14	0.12	3.7	2.44	0.68
15	< 0.10	1.7	1.23	0.38
16	< 0.10	0.62	0.49	0.17
17	< 0.10	0.19	0.16	0.06
18	< 0.10	0.05		

Table III. Observed Tracer Concentration vs. Concentrations Predicted for Various Solubilities — Experiment 3 (DEA)

STATION	C_{obs}	$C_{H' = 0}$	$C_{H' = 11}$	$C_{H' = 21}$
A 4	0.6 ppm			
5	0.7			
6	2.1		0.005 ppm	
7	1.4	0.1 ppm	0.042	0.018 ppm
8	1.3	1.5	0.225	0.084
9	2.1	6.9	0.628	0.221
10	2.5	20.2	0.898	0.304
11	1.0	6.9	0.628	0.221
12	0.6	1.5	0.225	0.084
13	< 0.1	0.1	0.042	0.018
14	< 0.1		0.005	
B 11	0.9	0.01	0.002	0.001
12	1.7	0.12	0.021	0.012
13	2.2	0.81	0.112	0.058
14	2.7	2.9	0.304	0.154
15	3.0	4.8	0.425	0.212
16	2.1	2.9	0.304	0.154
17	2.8	0.81	0.112	0.058
18	1.1	0.12	0.021	0.012
19	1.2	0.01	0.002	0.001

Table IV. Observed Tracer Concentration vs. Concentrations Predicted for Various Solubilities — Experiment 4 (DEA)

STATION	C_{obs}	$C_{H'=0}$	$C_{H=11}$	$C_{H'=21}$
A 6	0.19 ppm			
7	0.55			
8	0.62	0.03 ppm	0.014 ppm	0.006 ppm
9	0.96	1.2	0.357	0.153
10	1.3	20.7	1.4	0.478
11	1.0	1.2	0.357	0.153
12	0.85	0.03	0.014	0.006
13	0.56			
14	0.39			
B 12	0.21			
13	0.42			
14	1.3			
15	1.3	0.04	0.013	0.007
16	1.7	1.2	0.19	0.095
17	2.2	10.4	0.48	0.228
18	1.7	1.2	0.19	0.095
19	1.4	0.04	0.013	0.007
20	1.8			
21	1.3			
22	0.80			
23	0.45			

Table V. Observed Tracer Concentration vs. Concentrations Predicted for Various Solubilities — Experiment 5 (DEA)

STATION	C_{obs}	$C_{H'=0}$	$C_{H'=11}$	$C_{H'=21}$
A 8	0.17 ppm			
9	0.47			
10	0.37	0.04 ppm	0.02 ppm	0.01 ppm
11	0.43	0.80	0.24	0.10
12	0.57	10.7	0.70	0.25
13	0.32	0.80	0.24	0.10
14	0.48	0.04	0.02	0.01
15	0.17			
B 12	0.28			
13	0.41			
14	0.47			
15	0.47		0.001	0.001
16	0.47	0.08	0.019	0.01
17	----	0.76	0.112	0.054
18	0.63	5.03	0.207	0.097
19	0.56	0.76	0.112	0.054
20	0.43	0.08	0.019	0.010
21	0.43		0.001	0.001
22	0.42			

Refinement, if required, must await application to a variety of conditions so that deficiencies may be identified and corrected.

Acknowledgements

The authors gratefully acknowledge the financial support of the U. S. Army Research Office - Durham.

Literature Cited

1. Fuquay, J. J., Scavenging in Perspective In: "Precipitation Scavenging," pp 1-5, AEC Symposium Series 22, U. S. Atomic Energy Commission (1970).

2. Dana, M. T., J. M. Hales, W. G. N. Slinn, M. A. Wolf, "Natural Precipitation Washout of Sulfur Compounds From Plumes," Final Report to Environmental Protection Agency EPA-R3-73-047 (1973).

3. Hales, J. M., M. A. Wolf and M. T. Dana, AIChE Journal, (1973), 19, (2), pp 292-297.

4. Private correspondence from the U. S. Army Research Office, Durham.

5. Dahrgen, G., Analytical Chemicsty (1964), 36, (3), pp 596-599.

6. Dailey, B. P. and W. A. Felsing, J. Am. Chem. Soc., (1939), 61, (10), pp 2808-2809.

7. Perry, J. H., "Chemical Engineer's Handbook," McGraw-Hill, New York, (1966).

4

Introduction, Transport, and Fate of Persistent Pesticides in the Atmosphere

D. E. GLOTFELTY and J. H. CARO

Agricultural Chemicals Management Laboratory, Agricultural Environmental Quality Institute, Agricultural Research Service, USDA, Beltsville, Md. 20705

As has been often reported, worldwide production and use of organic pesticides grew explosively in the quarter century following the introduction of DDT during World War II. In the peak year of 1968, annual production in the United States reached more than 5.3×10^5 metric tons of active pesticide ingredients, including 2.64×10^5, 1.83×10^5, and 8.65×10^4 metric tons of insecticides, herbicides, and fungicides, respectively ([1]). The current domestic market includes more than 420 biologically active chemical compounds sold in over 6700 different formulations ([2]). These multitudinous products have specific, registered uses: they are applied primarily to agricultural crops, soils, livestock, forests, bodies of water, and the general environment, to assist in food and fiber production and to protect public health. They also are used commercially in food processing and protection of stored products, and are applied in the home and the garden.

Each of these purposeful applications provides a source for entry of the chemicals into the environment. Unintentional sources also exist, such as uncontrolled effluents from pesticide manufacturing and formulating plants, chemical spills, structural fires involving chemicals, and improper disposal of surplus chemicals or used containers. Once applied or released, the pesticides are transported from the original site by water or air. Considerable research has been conducted on the mechanisms and scope of transport and on the identification of terminal residues, yet broad gaps still exist in our understanding of the system. We examine here the current state of knowledge as it applies to the introduction and transport of pesticides in the atmosphere, present a unique approach to evaluating the relative transport potential of specific pesticides, discuss the processes by which the pesticides are removed from the atmosphere, and indicate the most fertile directions for future research.

Pesticide residues in air are of concern not only because of direct economic effects such as the deposition of illegal

42

residues on nontarget organisms and injury to susceptible crops, but perhaps more significantly, because of the possible long-range, subtle adverse effects on human health (3, 4). Of primary importance on a global scale are the persistent pesticides, notably the organochlorine insecticides as typified by DDT and its derivatives. Although use of DDT is now largely forbidden in many countries, including the United States, its long and intensive application over broad areas of land in recent years, its subsequent well-documented transport to untreated areas, and its continuing use in the developing countries for public health purposes have caused low levels of its residues to be ubiquitous on the surface of the earth as well as in the air (5). These residues are of particular concern because of their physicochemical properties, which are typical of those of the general class of chlorinated insecticides. They have low water solubility, high solubility in fats, and high resistance to chemical and biological decomposition. Consequently, they tend to persist and accumulate in living organisms and have demonstrated ability to concentrate in food chains, so that very low levels in the environment can cause serious adverse effects at the higher trophic levels. This behavior was not generally recognized for many years after they were first used, and much avoidable damage was wrought through their overuse and misuse. It is now imperative that we gain a thorough understanding of the mechanisms of transport of DDT and related pesticides away from the site of application and their eventual fate in the environment so that we may more judiciously use pesticides in the future.

Introduction of Pesticides into the Atmosphere

A brief, probably incomplete, inventory of some of the contributing sources of pesticides in air is given in Table I. With the exception of the agricultural inputs, we know very little about these sources. Each source does, however, release the pesticides into the atmosphere in a spectrum of particle sizes, so that considerations related to particle size, which are dealt with in the later section on aerial transport, are applicable to all sources. Owing to the large acreages and amounts of chemicals involved, agricultural use of pesticides deserves individual attention as a primary contributor to atmospheric contamination. Inputs from agriculture consist of (a) losses to the air during application of the pesticide, and (b) post-application losses, including wind erosion of contaminated surface dust and volatilization in molecular vapor form of pesticide residues on the surfaces of vegetation or soil (Table I). Discussions of each of these follow.

Application Inputs. Losses to the air during pesticide application often constitute 50% or more of the total pesticide used (6). Part of the material lost will return quickly to the surface a short distance downwind (drift); part will be deposited more slowly further downwind; and part will become airborne for a much longer time, diluting rapidly in the atmosphere, and returning to the surface very slowly. How much of the total loss falls into each of these categories depends on the particle size and the ambient weather conditions. For legal and economic reasons, drift has received much attention, but it is the pesticides that are subject to more protracted atmospheric transport that may ultimately be more significant.

Pesticides sprayed from helicopters or airplanes give substantially greater losses to the air than when applied as ground sprays, because the greater release height allows a longer fall time during which downwind transport and droplet evaporation can occur. On-target deposits of 5% or less of the nominal application have been reported for some aerial treatments (4). The greater release height is not necessarily due to inability of the aircraft to come sufficiently close to the surface, but can be caused by the air turbulence created in the wake of the aircraft.

Postapplication Inputs. Transport by wind erosion of contaminated surface dust to high altitudes and for long distances has been documented (7, 8), but it is more infrequent and less important than vaporization as a mechanism of pesticide input to the atmosphere. Not only highly volatile pesticides, but also the so-called "nonvolatile" pesticides such as the common organochlorine insecticides, can evaporate from soil, water, and vegetation surfaces in substantial quantities (9). Indeed, vaporization to the atmosphere may be a dominant pathway into the environment for compounds of low water solubility and long persistence, properties that the organochlorine pesticides possess. Evaporation begins when the pesticide is applied and continues as long as the compound is present.

The evaporation rate of a pesticide into air from soil, water, or vegetation is determined solely by the effective vapor pressure of the substance at the evaporating surface and the rate of diffusion through the layer of air immediately adjacent to the surface (10, 11). Air close to the evaporating surface does not move, and molecular diffusion across the laminar "skin" of air controls the volatilization rate. Volatilization increases if the depth of the laminar layer, normally 2 to 3 mm, decreases or if the concentration gradient across the layer increases. Many variables may affect volatilization by influencing the two basic factors. These variables include temperature, concentration of pesticide at the surface, degree of adsorption, type of formulation, rate of mass transport to the air-surface interface, supply of latent heat of vaporization, rate of air flow above the surface, and atmospheric turbulence.

Multilayered surface deposits such as foliar spray applications evaporate at a constant rate until so little remains that the surface is no longer uniformly covered (11). Evaporation rate from plant or soil surfaces may be very high shortly after application. In one field experiment, for example, a vapor concentration corresponding to saturation at the ambient temperature was observed for dieldrin in air 10 cm above a short stand of orchardgrass during the first 2 hours after foliar application; the field effectively behaved as a surface of pure dieldrin during this period (12).

The vaporization of a pesticide from soil is particularly influenced by its solubility in soil water and its degree of adsorption on soil particles. Soil water plays an extremely important role; it competes with the pesticide for adsorption sites and also carries solubilized residues to the surface (10, 11, 13). The rate of desorption of pesticide residues and their movement to the surface are controlled by temperature, pesticide concentration in the soil, and soil moisture content and bulk density. The long-term loss rate of organochlorine pesticides incorporated into soil is probably closely related to the mass flow of water to the soil surface (11). Differing loss rates from different soil types reflect the adsorption capacity of the soil, which is primarily related to its organic matter content. Loss is generally greater from a sandy soil of low organic matter content than from a high-organic soil. Incorporation into soil will sharply reduce volatilization losses; some relatively volatile herbicides lose activity rapidly by volatilization unless they are incorporated (14).

Volatilization of pesticides from vegetation surfaces has not been studied as extensively as that from soil, although it can be high for surface deposits. Volatilization is reduced by solubilization of the surface pesticides in plant waxes and oils, penetration to internal portions of the plant, and translocation. The formulation and method of application may be important (15), as are any chemical changes in the pesticide caused by the plant or by sunlight. The effect of plant transpiration on pesticide volatilization is unclear, although water evaporation from plant leaves is known to change the energy budget. Plants with adequate moisture are relatively cool because the incoming solar radiation is dissipated as latent heat of evaporation; water evaporation is obviously restricted under dry conditions and the plants become warmer through insolation (receipt of incoming solar radiation). Cooler plant surfaces may restrict pesticide volatilization.

The gross flux of pesticide passing into the atmosphere by volatilization from a treated area can be estimated in the field, using accepted micrometeorological techniques (16). A number of specific procedures are available, all of which require measurement of the pesticide concentration in air at several heights above the field. The basic technique depends upon calculation of an appropriate diffusivity coefficient, which is then related to the pesticide concentration gradient to give a value for pesticide flux. The validity of the numbers so obtained depends on how closely the field conditions approach the theoretically ideal, horizontally uniform, two-dimensional surface. Aerodynamic properties of a given terrain may make the method impractical in some situations, but successful measurements have been made under field conditions (12, 17, 18).

The results of these experiments, conducted with the insecticides dieldrin and heptachlor, showed that postapplication volatilization is a significant pathway of loss even when the chemicals are incorporated into the soil. In a corn field, volatilization of soil-incorporated dieldrin and heptachlor during the crop season totaled 3.6 and 6.8%, respectively, of the nominal application (18). These losses were larger than those by other measured pathways such as sediment movement, runoff water transport, and crop uptake (19). This is significant, because volatilization is uncontrolled, continuous transport, whereas sediment transport, the other major route of loss, can be largely controlled by common agricultural practices.

Heptachlor and dieldrin applied as a surface spray to a field of orchardgrass volatilized from the vegetation much more rapidly than when they were incorporated into the soil (12). The data showed that 90% of the residues of both compounds initially on the grass volatilized within 3 weeks. Whether incorporated or left on the surface, the pesticide flux into the air showed very strong diurnal variation, with the maximum flux occurring around solar noon. An obvious reason for this is the greater surface heating caused by the more direct insolation at noon. This, in turn, provides latent heat of vaporization, increases thermal instability and turbulence in the air, and increases the upward rate of water movement to the evaporating surface, bringing with it associated pesticide by bulk movement. In sum, there is little doubt that aerial transport from treated lands is a major mechanism of dissipation of the organochlorine insecticides into the environment.

Evaporation of a pesticide from a water surface depends upon the vapor pressure of the compound, its solubility, and the amount of the residue that is in true solution. The very low water solubility of organochlorine insecticides yields only very dilute, ideal solutions for which Henry's Law is obeyed. Henry's Law specifies that the vapor pressure over the solution is proportional to both the vapor pressure of the pure compound at that temperature and the relative saturation of the solution. Volatilization rates from water can be very high, especially for low solubility compounds (20). In addition to being in true

aqueous solution, pesticide residues in natural waters may be
associated with suspended particulate matter, incorporated into
living or dead organisms, or preferentially dissolved in organic
films. All of these serve to lower the volatilization rate,
which may then depend on the rate at which the residue is added
to or removed from true solution, or on the rate of mixing
wherever the pesticide in the surface layer becomes depleted.
Bottom sediments, which often serve as a reservoir of pesticide
residues, may release the residues into solution very slowly,
so that volatilization from the water body may reflect the slow
desorption and mixing processes. To counterbalance these
volatilization-reducing processes, an enhancement process may also
take place. The lipid-soluble, hydrophobic pesticides may preferentially dissolve in the natural organic surface films associated
with all water bodies. They then may become airborne by the
action of spray or bursting bubbles at the surface, at a rate
exceeding that predicted by Henry's Law for the small amounts
in true solution (21, 22).

Transport of Pesticides in the Atmosphere

Two spatial scales of transport from the point of pesticide
application exist: local, which includes distances up to about
15 miles, and long range, which ultimately includes global transport. Drifts of sprays and dusts caused by local air transport
can produce substantial application losses and result in adverse
biological effects on nontarget organisms (including man) and
economic losses from damage to crops and ornamental plants.
These, as well as the deposition of illegal residues, have resulted in numerous lawsuits against pesticide applicators and
manufacturers (4, 23, 24). Global-scale pesticide transport by
air is less well understood because the tremendous dilution
produces very low concentrations and because the diverse input
and removal mechanisms are not clearly defined. Obviously, such
transport does occur, since DDT residues have been found in many
remote environments (6, 25, 26, 27). Increased knowledge of the
transport process and the significance of transported residues
has important implications for future decisions concerning pesticide use.

The extent to which a pesticide is dispersed in the atmosphere depends primarily on its particle size and its constitution.
In studying atmospheric transport, the basic questions that must
be answered include: (1) What is the particle size? (2) Is the
airborne particle likely to change in size? (3) Will the chemical
nature of the airborne pesticide change? For purposes of orderly
discussion, it is convenient to classify the processes that relate
to particle size, particle size change, and chemical change as
primary, secondary, and tertiary, respectively, as in Table II.
Conceptually, a particle is considered to include spray droplets,
dust particles, atmospheric aerosol, wind-blown soil particles,

Table I. Principal Sources of Pesticides in Air

Activity	Sources
Agriculture	Application losses, drift, postapplication volatilization and wind erosion
Public health	Spraying for control of insect vectors of disease
Industrial	Pesticide manufacture and formulation, effluents, fumes, vapor, dust
Commercial	Mothproofing, food processing
Home and garden	Local application
Accidents	Spills, improper disposal

Table II. Processes Affecting the Transport of Pesticides in Air

Classification	Processes
Primary	Atmospheric: advection, convection, turbulence, eddy diffusion, frontal lifting Gravitational: fallout, sedimentation Small particle: impaction, deposition, molecular diffusion
Secondary	Evaporation and condensation Nucleation Coagulation Sorption Solution-vapor equilibrium
Tertiary	Chemical reactions: oxidation, hydrolysis, catalytic decomposition Photochemical reactions: direct or indirect, isomerization, oxidation

and the like, and for some of the physical processes discussed
below, the behavior of molecular vapor is indistinguishable from
that of other small particles.

Particle Size of Inputs. The particle sizes of pesticides
that become airborne during application of the chemical vary with
the formulation and application equipment. Granular formulations
consist of particles that are 200 to 600 µm in diameter and have
no direct potential for dispersion. Dusts are screened during
manufacturing to 80-90% less than 30 µm and a particle number
median diameter of 1-10 µm (28). These fine particles have high
transport potential, an undesirable attribute that has contributed
to decreased use of dust formulations despite their efficient
coverage of plant surfaces (29). Sprays, by far the most widely
used formulations, may consist of a water or oil base in which
the pesticide is dispersed, an emulsifiable liquid, or a wettable
powder in which the active chemical is on the surface of clay
particles. All common spray equipment produces a spectrum of
particle sizes ranging from 1 to over 400 µm. For a sprayer
set for 200-µm droplets, up to 8% of the spray volume is in
droplets finer than 200 µm (4). Thickened sprays such as water-
in-oil (invert) emulsions may produce larger spray drop averages.
Small drops can not be eliminated with present equipment, merely
reduced in number (29).

The particle sizes of postapplication inputs include molecular vapor from volatilizing pesticides and a wide range of
particle sizes from wind-eroded dust. Long-range aerial transport
of pesticides on dust is probably associated with particles finer
than 10 µm in diameter.

Transport and Particle Size. Particles descend through the
atmosphere under the influence of gravity by well-known physical
laws that govern the resistance to motion of particles moving in
viscous media. The descent is described by the Stokes equation,
commonly found in physics texts. Larger particles, as is evident,
tend to settle to the surface rather rapidly, but as particle
size decreases, the effect of gravity becomes less owing to
buoyancy, viscous forces, and the random motion of the air in
which the particles are suspended (3, 30, 31). The transport
potential then depends on the balance achieved between particle
size, gravity, and the strength of these random motions. The
physical forces involved are shown in Table II as primary
processes.

Particles larger than 100 µm in diameter fall relatively
rapidly and are largely unaffected by air turbulence. Smaller
particles may have a positive fall velocity, but their motion
is dependent upon the strength of the turbulence. Particles of
10 µm or less are dispersed primarily by turbulence under all
meteorological conditions (31). Molecular vapor is returned to
the surface with difficulty, and long-range transport is indeed
very probable. Estimates of atmospheric residence time based on

rates of removal of surface dust or atmospheric aerosol are not
adequate in describing the fate of molecular vapor in air.
Estimates of the amount of time that molecular vapor may persist
in air are largely speculative and depend upon the rate of
secondary and tertiary processes (Table II), concerning which
little experimental information is available.

Transport and Particle Size Change. Once pesticide particles
become airborne, they may change in size. The physical forces
involved are shown in Table II as secondary processes. The
particles can become either larger or smaller. For example, a
spray droplet can evaporate to a smaller size, or molecular
vapor may sorb to an aerocolloidal particle or partition into a
raindrop, so that the resulting particle is larger. The transport potential, of course, changes with the change in particle
size.

Transport and Chemical Change. The tertiary processes listed
in Table II are those that change the chemical nature of the
airborne pesticide. These processes are doubly important, in
that they not only profoundly affect the transport potential but
also change the analytical and toxicological properties of the
pesticide. Unfortunately, we know the least about these processes
as they operate in the atmosphere, largely because of uncertainty
about the physical state of airborne pesticides and about how
long one may reasonably expect a pesticide to remain airborne.
Considering that the atmosphere is a highly reactive medium and
that many pesticides undergo sunlight-induced photochemical
reactions under laboratory conditions, the effect of tertiary
changes cannot be neglected when describing pesticide transport
by air.

Meteorological Conditions Affecting Transport. The mixing
characteristics of the atmosphere are governed in a major way by
its vertical temperature gradient, but also depend upon surface
roughness, which can create turbulence due to wind shear. The
turbulent mixing process is a complicated structure of eddies, a
random motion that generally follows the mathematical laws of
diffusion, but with a diffusion coefficient 10^2 to 10^4 times
greater than that of molecular diffusion. Since air flow above
the surface is practically never laminar, molecular diffusion
can usually be neglected, except very close to an evaporating or
adsorbing surface. The diffusion coefficient for a collection
of particles in the atmosphere may actually show enormous
variations, ranging from 0.2 cm^2/sec for molecular diffusion
to 10^{11} cm^2/sec for large-scale cyclonic storms (30).
The magnitude of eddy diffusion may change diurnally, owing
to differences between surface and air temperature that are
created by solar heating and nocturnal radiational cooling of

the surface. Insolation raises the surface temperature so that
the relatively cool overlying air is most unstable in the early
afternoon. The temperature lapse rate, and thereby turbulent
mixing, is much less at night for a given wind speed, and nocturnal
inversions (warm air overlying cool air) are common. Crop irri-
gation modifies these temperature conditions by transforming more
of the incoming solar energy into latent heat by evapotranspira-
tion, and the air is not as unstable over an irrigated crop as
over dry land.

Surface roughness, also a factor in creating air turbulence,
may depend on the crop species and wind speed. For example, a
tall pliant crop such as wheat may bend in the breeze, sealing
the top of the crop and preventing wind penetration close to the
surface. A relatively rigid crop, such as corn, may allow better
wind penetration and, thereby, increased transport away from the
surface (24).

Air moves pesticide away from the treated area by a combination of horizontal (advection) and vertical transport
(convection). The extent of each of these is determined primarily
by air turbulence. Advection is favored by low turbulence caused
by thermal stability and low wind shear. Under these conditions,
which are typical of inversion situations, long periods of low-
lying horizontal transport occur. Pesticide concentration along
the downwind surface remains high and local transport into adjacent areas with rapid return to the surface, primarily typified
by fine particle "drift" from application operations, is favored
(24, 31). Windy conditions, as most farmers know, also increase
drift, because they favor increased advection of larger particles
that normally settle rapidly to the surface. The application of
fine particles, dust or spray, under convective conditions of
thermal instability and moderate winds may lessen local drift
problems by rapidly diluting airborne residues and maintaining
low concentrations at the downwind surface. However, the increased vertical component increases the probability of long-range
transport and, although surface residues immediately downwind may
be reduced, application losses may increase. If the convected
residues eventually return to the surface, the drift has, in
reality, only been diluted.

Removal of Pesticides from the Atmosphere

Pesticides are removed from the atmosphere by four basic competitive mechanisms: dry deposition, precipitation scavenging,
chemical degradation, and photochemical degradation. The
processes that affect atmospheric transport of pesticides (Table
II) also affect their removal, since removal, as does transport,
depends primarily on the size and the nature of the pesticide
particle. Discussions of the individual factors influencing the
removal process follow.

Pesticide-Aerosol Interactions. The atmosphere contains an aerosol consisting of an enormous variety of minute, suspended solid and liquid particles that continually circulate throughout the air space. Although considerable variation can occur, the number density of these particles is generally about 1 to 10 per cc of air for the important particle size range between 0.1 and 10 μm (32, 33). Pesticide particles, including molecular vapor, that are introduced into this aerosol may interact with it, resulting in effective growth in pesticide particle size and thereby hastening removal of the pesticide from the general circulation. The large surface-to-volume ratio of atmospheric aerosols, their demonstrated ability to absorb or react with trace gases, and their relatively short residence time in the atmosphere have led to speculation that adsorption of pesticide vapor onto aerosol particles may be a major removal mechanism (31, 34). The presence of particulate pesticides in the atmosphere has been confirmed (26, 27, 35), but conclusive evidence is lacking that they form by pesticide vapor-aerosol interaction as distinguished from wind-eroded dust or particulates from pesticide application operations. The importance of pesticide-aerosol interaction as a removal mechanism could be better evaluated if more information were available on the probability of collision, the adhesive force, and the composition and properties of the aerosol (36, 37).

In addition to increasing the effective particle size, aerosol adsorption could also promote important tertiary changes such as catalyzed chemical reactions or sensitized photochemical conversions. Strong adsorption onto an aerosol may even change the wavelength of maximum absorption in direct photochemical reactions (38). However, effects on tertiary processes in the atmosphere have not been investigated directly because of the inherent experimental complexities.

The fact that all atmospheric aerosols contain organic matter and that this organic matter is known to react chemically constitutes evidence that aerosols might well absorb organic pesticides and control their fate. The source and composition of some of the organic matter in aerosols have been identified. For example, maritime aerosols contain organic matter having a composition very similar to that of naturally occurring oil slicks on the surface of the ocean (21, 22). Aerosols over land contain a variety of reactive organic compounds, especially terpenes, that are released by vegetation (39). Aerosols over urban areas contain organic matter that is partially in the form of combustion-derived polynuclear aromatic hydrocarbons (32), which have physicochemical properties similar to those of many organic pesticides. They also have been shown to contain detectable quantities of DDT (35), so that pesticide association with aerosol organic matter is documented. Reactions in aerosols generally involve oxygenations (39); therefore, the products of any pesticide decomposition that would occur would likely be polar, water-soluble compounds. Because present analytical methods for organochlorine insecticides are based primarily on the consistent lipid solubility of the compounds, water-soluble products will likely be overlooked. This could explain some of the unaccounted-for losses of DDT in the environment.

Photochemical Degradation. The extensive recent literature on the photochemistry of pesticides has shown that many pesticides react photochemically in the laboratory, often yielding complex mixtures of products. Frequently, the products isolated in the laboratory can indeed be found in the environment, and sunlight-induced photochemical reactions have been demonstrated in the field. However, the importance of photochemical reactions in the atmosphere has not yet been validly assessed, primarily because basic information is lacking. For example, a photochemical reaction can be important only if it is fast enough to compete with removal processes that return pesticides to the surface. This requires knowledge of the mean atmospheric residence time of the pesticide and of the quantitative relation between photochemical reaction velocity and the intensity and wavelength of sunlight in the lower atmosphere. These quantities are uncertain because they depend upon the composition of the particles containing the pesticide.

Thus far, work on the photochemical degradation of pesticides has yielded information important to the understanding of possible atmospheric reactions. Laboratory studies have shown, for example, that intramolecular eliminations and isomerizations tend to dominate when pesticides in molecular vapor form are irradiated, whereas other reactions, such as oxidation or condensation, can occur with pesticides dissolved in water or organic solvents. A pesticide strongly adsorbed to a surface can exhibit dramatic shifts in its absorption spectrum (38). A photosensitizer can strongly affect reaction rate, so that the composition of the atmospheric particles containing the pesticide is very important (38, 40).

The photochemical reactivity of many of the organochlorine insecticides has been demonstrated in the laboratory. The cyclodiene insecticides, such as dieldrin, heptachlor, or aldrin, typically undergo isomerization or dechlorination. Although dieldrin shows negligible absorption at solar wavelengths that reach the surface of the earth, it readily reacts under natural or artificial sunlight, even in the absence of any apparent photosensitizer. Pure molecular vapor of dieldrin in air or inert gas yields only a single product, photodieldrin, resulting from isomerization (41, 42). Photodieldrin has been found in the environment, forming on vegetation and soil surfaces, in natural waters, and presumably in air. It is more toxic than dieldrin and much less volatile under field conditions. If dieldrin occurs in polluted atmospheres containing ozone or

nitrogen oxides, other more polar products of irradiation may
form in addition to photodieldrin (43). DDT decomposes slowly
under direct photolysis at solar wavelengths, converting to DDD
by an oxygen-inhibited process. However, like other chlorinated
compounds, DDT also undergoes a photochemical reaction initiated
by aromatic amines that is not oxygen inhibited, but yields
oxygenated products (44). DDT has even been shown to convert to
polychlorinated biphenyl (PCB) under long irradiation times (45).
As mentioned, the ultimate importance of these photochemical
reactions in the degradation of airborne residues cannot now be
ascertained.

Dry Deposition. The deposition of particles from the atmosphere onto exposed surfaces in the absence of rain or snowfall,
so-called "dry" deposition, occurs as a long-term process for
virtually all airborne materials. It will continue as long as
the air mass containing the material remains in contact with the
ground. Thus, surface deposition may be a very effective
mechanism for removing airborne residues even though particle fall
velocity is negligible (31). Empirical observations allow the
rate of particle removal by surface deposition to be quantified
without actually specifying the processes involved. The quanti-
fication depends upon the proportionality between the rate of
surface accumulations per unit area and the surface airborne
concentration. This ratio of flux to concentration has dimensional
units of velocity, and this deposition "velocity" can be used to
calculate depletion rates of a collection of airborne particles.
Deposition velocity depends primarily upon particle transport and
hence is a function basically of particle size rather than of the
composition of the particle. Molecular vapor is an exception,
in that deposition velocities, which may vary widely, depend upon
the reactivity of the molecule.
    The deposition velocity of 100-μm particles as measured by
deposits on vegetation approximates 50 cm/sec, and decreases with
decreasing particle size to about 0.03 cm/sec for nonreactive
molecular vapor having a particle diameter of about 0.001 μm.
By contrast, a reactive vapor such as molecular iodine moves to
the surface at a diffusion-controlled rate of about 2 cm/sec.
The organochlorine insecticides exhibit deposition velocities
approaching that of nonreactive molecular vapor, indicating that
attachment at the surface, rather than diffusion to the surface,
is the rate-controlling step (46).
    Large particle deposition velocity approximates the terminal
fall velocity as defined by the Stokes equation. The deposition
velocity of smaller particles exceeds their terminal fall velocity, reflecting the increased importance of turbulent mixing
in bringing small particles to the surface. Except for reactive
vapor, the deposition velocity is approximately constant for

particles below 1 μm. This is because both very fine particles
and unreactive molecular vapor are transported across the laminar
layer surrounding exposed vegetation surfaces by Brownian motion,
with impaction becoming less important as particle inertia
decreases.

In using deposition velocity data, it is convenient to specify
a plume height or mixing height $\underline{H}$ throughout which rapid,
uniform mixing is assumed, and then to integrate the relation

$$- H/N \frac{dN}{dt} = V_{deposition}$$

where $\underline{N}$ is the total number of particles in the column of air
over a unit area of surface up to height $\underline{H}$. The calculated
times required for 50% depletion of various particle sizes by
deposition on vegetation are given in Table III for a 2-km
mixing height. The times shown agree in magnitude with those
reported in the literature (46).

Precipitation Scavenging. The inclusion of airborne pesticide residues in falling rain or snow is competitive with dry
deposition as a removal mechanism. A spray droplet or dust
particle that is in the atmosphere for an appreciable period of
time as a discernible entity may serve as a nucleating particle
in the growth of a raindrop and would soon fall to earth. The
importance of precipitation scavenging as a removal process
relative to that of dry deposition is greater for fine particles,
because coarse particles fall out more rapidly and are not
available for scavenging for as long a period of time.

Rainfall may also remove molecular vapor. The solubilization
of molecular vapor in a raindrop is essentially a partitioning
process in which a given concentration of gaseous pesticide
establishes an equilibrium concentration in solution. Since
dilute solutions are involved, it is very likely that they will
be ideal and Henry's law will be obeyed; that is, the ratio of
partial pressure to equilibrium solution concentration is a
constant, K_H, called Henry's constant. Ideality has been confirmed for the organochlorine insecticides DDT, dieldrin, and
lindane (46). The reasonably good agreement of the washout
ratios for these compounds as measured in the laboratory and in
the field has been purported to show that these chemicals exist
in the atmosphere essentially as unassociated molecular vapor
(46).

Atmospheric Residence Times and Transport Potentials

How long residues of persistent pesticides may exist in the
atmosphere is largely a matter of speculation because of the many
interrelated processes involved and the difficulties inherent in
making quantitative measurements. As a result, published esti-

mates have been widely divergent; for example, the atmospheric
residence time for DDT has been estimated variously as less than
2 weeks (31) and about 4 years (5). Much of the discrepancy is
based on whether the airborne pesticide is considered as being
chiefly in particulate form or in molecular vapor. Large particles are removed rather effectively by surface deposition or
rainout and the mean atmospheric residence time will be fairly
short; molecular vapor, on the other hand, can remain airborne
for extended periods. Evidence is now accumulating that chlorinated organics exist in the atmosphere largely as molecular vapor
(25, 46, 48), so that longer residence times appear to be more
realistic and the relative potentials for long-term transport of
individual compounds assume greater importance.

Deposition velocity data can be used, with a number of
simplifying assumptions, to predict atmospheric residence times,
provided that the long-term input and removal of the pesticides
are essentially equal. Such an approach may have some validity
for the steady state reached between latent vaporization and
deposition of molecular species. On this basis, it is predicted
that organochlorine insecticides dispersed as molecular vapor
throughout the troposphere (atmosphere to a height of 12 km)
would be depleted by vegetation by about 4% per week, giving a
mean residence time of about 25 weeks (46).

Transport Potentials. The amounts of several common pesticides found in remote environments have been found to correlate,
within the limits of the chemical stability of the compounds,
with their individual Henry's constants, which reflect the
partitioning of the chemicals between air and water. A ranking
for 19 pesticides and related materials based on this concept
is shown in Table IV. In the table, the chemicals are ranked
according to decreasing values of an apparent Henry's constant,
the ratio of saturation vapor pressure to aqueous solubility of
the pure compounds, and thereby according to decreasing transport
potentials. For the most part, the data shown are published
values of the properties. Where more than one value was found,
an average or generally accepted value was taken. Where values
were not available, we assigned a reasonable estimate. The
available information has obvious shortcomings; reported values
sometimes differ by as much as an order of magnitude, which
could change the ranking order in the table substantially. More
extensive and more reliable vapor pressure and solubility data
are acutely needed for these pesticidal compounds.

In general, a combination of high vapor pressure and low
aqueous solubility (high K_H) describes a compound with high
evaporation potential from concentrated aqueous solution and
resistance to removal of vapor from the atmosphere, to the
extent that this is controlled by transport across an air-water
interface. A high K_H also signifies that rainfall washout is

Table III.   Times for 50% Depletion of Particles from the
Atmosphere by Dry Deposition on Vegetation

Particle Diameter, μm	Terminal Fall Velocity, cm/sec	Empirical Deposition Velocity, cm/sec	50% Depletion Time for 2-Km Mixing Height, hrs.
100	50	50	(0.8)[a]
10	1	2	20
1	$<10^{-2}$	0.15	260
0.1	10^{-3}	0.06	640
0.001	--	0.03	1280

[a] Particles generally will not mix uniformly throughout the
2-km height because of high terminal fall velocity.

Table IV.   Apparent Henry's Constants of Pesticides and Related
Materials

Compound	Vapor Pressure at 20–25°C mm Hg	Water Solubility at 20–25°C $\mu g/1$	Log Apparent K_H
Toxaphene	3×10^{-1}	10^2 to 10^3	-2.5 to -3.5
Aroclor 1242-1260	4×10^{-4} to 4×10^{-5}	2.7 to 240	-4.8 to -5.8
Heptachlor	3×10^{-4}	56	-5.3
2,4-D Esters	10^{-3} to 10^{-4}	(10^3)[a]	-6.0 to -7.0
γ-Chlordane	10^{-4} to 10^{-5}	(10^2)	-6.0 to -7.0
Aldrin	2.3×10^{-5}	27	-6.1
Trifluralin	10^{-4}	580	-6.8
p,p´-DDT	1.5×10^{-7}	1.2	-6.9
o,p´-DDT[b]	5.5×10^{-6}	85	-7.2
p,p´-DDE[b]	6.5×10^{-6}	120	-7.3
EPTC	1.6×10^{-2}	3.8×10^5	-7.4
Dieldrin	2.8×10^{-6}	140	-7.7
Heptachlor Epoxide	$(10^{-4}$ to $10^{-6})$	350	-6.5 to -8.5
Diazinon	2.8×10^{-4}	4×10^4	-8.2
Lindane	3.3×10^{-5}	7×10^3	-8.3
Endrin	2×10^{-7}	2×10^2	-9.0
Parathion	2.3×10^{-5}	2.4×10^4	-9.0
Carbofuran	(10^{-5})	2.5×10^5	-10.4
2,4-D Salts	$\sim 10^{-10}$	Soluble	--

[a] Values in brackets are estimated
[b] Data at 30°C.

not a major factor in the removal of molecular vapor of that compound from the atmosphere. For relatively nonreactive compounds such as the organochlorine insecticides, for example, those having K_H values greater than about 10^{-8} (Table IV) appear to be more effectively removed by dry deposition on vegetation surfaces. For lindane, which has a K_H value slightly below 10^{-8}, the contributions of dry deposition and rainout are predicted to be about equal (46). Endrin, the lowest ranked organochlorine insecticide, would presumably be removed primarily by rainfall washout.

Some examples serve to illustrate the conformance of the ranking to known environmental behavior. Quantitites of Aroclors (PCB's) present in the marine atmosphere are about 20-fold higher than those of DDT (25), and they are found near the top of the list. The fairly high transport potential indicated for chlordane predicts its presence in remote environments, since it is chemically stable. This has recently been confirmed by measurements over the Atlantic (25). Parathion is predicted to have a low transport potential, and this has been confirmed in an experiment in which only small quantities of parathion were found to be transported beyond the area of direct fallout of spray droplets (49). The K_H values in Table IV predict that DDT and its congeners should have about the same transport potential despite the fact that their vapor pressures are significantly different. Also, aldrin and heptachlor are shown to have a high transport potential, but they are chemically reactive and probably undergo a tertiary process leading to dieldrin and heptachlor epoxide, which have lower transport potentials.

Henry's constant is also helpful in gaining information on spray drift. As an aqueous spray drop evaporates, the volume becomes small, the pesticide in the drop becomes concentrated, and compounds with high K_H values tend to establish high vapor pressure. Thus, 2,4-D esters are transported downwind mainly as vapor, whereas salts, such as the dimethylamine salt, are transported almost exclusively as spray droplets (4). The drift potential of the former is much greater than that of the latter and the difference increases rapidly with downwind distance.

Research Needs

Many of the unanswered questions of fundamental importance within the very broad subject of pesticides in the atmosphere can be addressed directly, using present-day equipment and techniques. Other questions are more difficult and require more basic research. Some of the conclusions drawn here, although based on the best available information, must be regarded as tentative because of the unavailability of hard data. We believe that long-range atmospheric transport of persistent pesticides is important, although other opinions have been expressed (50).

Many research needs are clearly indicated in the text. With respect to long-range transport, elucidation of the role of the secondary and tertiary processes listed in Table II is particularly important. The list of research needs below, grouped for convenience into four categories, is not meant to be comprehensive, but constitutes major areas where research may well be fruitful.

General Needs. 1. Present methods of sampling the atmosphere are imperfect. Research can be directed to the continuing development of sampling methods in which small quantities of pesticides are quantitatively trapped from large volumes of air and kept chemically stable until they can be analyzed (51, 52, 53). 2. Analytical methods are needed that accurately distinguish between particulate matter and molecular vapor. Present particulate traps are of uncertain reliability because pesticides adsorbed on trapped particles may revolatilize during the continuing sample collection. 3. More comprehensive global monitoring of atmospheric pesticides is needed so that practicable goals may be set and standards defined with respect to air quality (4). 4. To aid in evaluating the ultimate significance of atmospheric pesticides, in-depth studies are needed of their acute and chronic effects on public health.

Needs Relating to Introduction. 1. For spray application of pesticides, further development is needed of techniques whereby spray drop size patterns may be adjusted to eliminate small drops that drift and vaporize. Present methods permit some reduction in number of small drops, but not elimination. 2. Study is needed to identify meteorological conditions that minimize total input to the atmosphere from application losses. 3. Work directed to improvements in pesticide formulations and application techniques to minimize losses to the atmosphere is needed, such as the continued search for systemic pesticides that are effective in granular formulation.

Needs Relating to Transport. 1. There is a large gap in our knowledge of the biosphere with respect to the relative contributions of the possible pathways, including aerial transport, to the worldwide distribution of persistent pesticide residues. Definitive studies are needed, but experimental difficulties are great. 2. Long-term and long-distance transport of airborne pesticides are less well understood than local transport. Mathematical models describing distant transport exist (5, 54), but additional information is needed to refine the models by testing the assumptions and predictions made. 3. The mathematical description of turbulent diffusion should be refined so that the transport of a contaminant in the atmosphere can be defined more precisely. Present methods are basically statistical and describe behavior only under a given set of meteorological conditions (30, 34).

Needs Relating to Fate. 1. The total process controlling the ultimate fate of airborne pesticide residues is only partly understood. All work directed to understanding the complex relationships involved, including much of the specific research cited below, is greatly needed. 2. The final resting place of pesticides released into the environment is regarded by many investigators as being the ocean abyss. Studies are needed of pesticide diffusion to and from the ocean surface, the transport properties of the surface microlayer, and the mechanism of transfer of residues to the ocean deep, to clarify the partitioning of pesticides between atmosphere and hydrosphere. 3. Similarly, investigations are needed of the exchange and the attainment of equilibrium in the concurrent processes of deposition and revolatilization of pesticide residues on land surfaces. 4. Better definition is needed of the actual residence times of pesticides in the atmosphere. 5. The influence of climate on the degradation, transformation, and removal of pesticides in air requires detailed study (29). 6. Quantitative studies are needed of the rates of removal of airborne pesticides by precipitation scavenging. 7. Research is also needed on washout ratios (ratio of rain-to-air concentration at a given sampling point) to yield fundamental information on the physical forms of the pesticides in the atmosphere (46). 8. Further studies of photolytic transformations in the atmosphere are greatly needed, especially field studies to complement the research which to date has been conducted largely in the laboratory. 9. A method is needed to permit differentiation of photolytic products formed in the air from those formed on the surface and volatilized. This is important in describing the eventual fate of atmospheric pesticides. 10. There always is an unaccounted-for shortage whenever attempts are made to measure a material balance for persistent pesticides, especially DDT, in the biosphere. Techniques are needed for assay of unsuspected alteration products, such as water-soluble derivatives, that may account for part of the shortage.

Literature Cited

1. The Pesticide Review, 1973. Agr. Stabil. Conserv. Serv., USDA, Washington, D.C., 1974.
2. Pesticide Handbook-Entoma, 25th Ed., Entomol. Soc. Amer., College Park, Md., 1974.
3. Van Middelem, C. H., Advan. Chem. Ser. (1966) 60, 245-48.
4. Grover, R., Maybank, J., Yoshida, K., Plimmer, J. R. (1973). 66th Ann. Mtg., Air Pollut. Contr. Assoc., Chicago, Ill.

5. Woodwell, G. M., Craig, P. P., Johnson, H. A., Science (1971) 174, 1101-07.
6. Westlake, W. E., Gunther, F. A., Advan. Chem. Ser. (1966) 60, 110-21.
7. Cohen, J. M., Pinkerton, C., Advan. Chem. Ser. (1966) 60, 163-76.
8. Risebrough, R. W., Huggett, R. J., Griffin, J. J., Goldberg, E. D. Science (1968) 159, 1233-35.
9. Pionke, H. B., Chesters, G., J. Environ. Qual. (1973) 2, 29-45.
10. Hartley, G. S., Advan. Chem. Ser. (1969) 86, 115-34.
11. Spencer, W. F., Cliath, M. M., Symp. "Environmental Dynamics of Pesticides," 167th Nat. Mtg., Amer. Chem. Soc., Los Angeles, Calif., Apr. 1974.
12. Taylor, A. W., USDA, Beltsville, Md., unpublished data.
13. Spencer, W. F., Cliath, M. M., J. Agr. Food Chem. (1974) 22, 987-91.
14. Parochetti, J. V., Warren, G. F., Weeds (1966) 14, 281-85.
15. Phillips, F. T., Chem. Ind. (London) (1974) No. 2, 193-97.
16. Parmele, L. H., Lemon, E. R., Taylor, A. W., Water, Air, Soil Pollut. (1972) 1, 433-51.
17. Caro, J. H., Taylor, A. W., Lemon, E. R., in "Proc. Symp. Identification and Measurement of Environmental Pollutants," 72-77, Nat. Res. Council Canada, Ottawa, 1972.
18. Taylor, A. W., Glotfelty, D. E., Glass, B. L., Freeman, H. P., 163rd Nat. Mtg., Amer. Chem. Soc., Boston, Mass., Apr. 1972.
19. Caro, J. H., Taylor, A. W., J. Agr. Food Chem. (1971) 19, 379-84.
20. Mackay, D., Wolkoff, A. W., Environ. Sci. Technol. (1973) 7, 611-14.
21. MacIntyre, F., Sci. Am. (1974) 230 (5), 62-77.
22. Barger, W. R., Garrett, W. D., J. Geophys. Res. (1970) 75, 4561-66.
23. Jegier, Z., Ann. N.Y. Acad. Sci. (1969) 160, 143-54.
24. Riley, J. A., Giles, W. L., Agr. Meteorol. (1965) 2, 225-45.
25. Bidleman, T. F., Olney, C. E., Science (1974) 183, 519-21.
26. Prospero, J. M., Seba, D. B., Atmos. Environ. (1972), 6, 363-64.
27. Stanley, C. W., Barney, J. E., Helton, M. E., Yobs, A. R., Environ. Sci. Technol. (1971) 5, 430-35.
28. Yates, W. E., Akesson, N. B., "Pesticide Formulations," p. 284, Marcel Dekker, Inc., New York, 1973.
29. Scotton, J. W., "Atmospheric Transport of Pesticide Aerosols," U.S. Dept. Health, Educ., Welfare, Pub. Health Serv., Washington, D.C., 1965.
30. Eisenbud, M., "Environmental Radioactivity," 2nd ed., Academic Press, New York, 1973.
31. Pooler, F., Jr., "Atmospheric Transport and Dispersion of Pesticides," COM-72-10454, Nat. Tech. Inform. Serv., U.S. Dept. Commerce, 1972.

32. Corn, M., Montgomery, T. L., Esmen, N. A., Environ. Sci.
     Technol. (1971) 5, 155-58.
33. Blifford, I. H., J. Geophys. Res. (1970) 75, 3099-103.
34. Pitter, W. L., Baum, E. J., Symp. "Environmental Dynamics
     of Pesticides," 167th Nat. Mtg., Amer. Chem. Soc.,
     Los Angeles, Calif., Apr. 1974.
35. Antommaria, P., Corn, M., DeMaio, L., Science (1965), 150,
     1476-77.
36. Judeikis, H. S., Siegel, S., Atmos. Environ. (1973) 7,
     619-31.
37. Hidy, G. M., Brock, J. R., "The Dynamics of Aerocolloidal
     Systems," 43-46, Pergamon Press, New York, 1970.
38. Plimmer, J. R., in "Fate of Pesticides in the Environment,"
     47-76, A. S. Tahori, Ed., Gordon and Breach, New York, 1972.
39. Cadle, R. D., in "Aerosols and Atmospheric Chemistry,"
     141-47, G. M. Hidy, Ed., Academic Press, New York, 1972.
40. Crosby, D. G., Moilanen, K. W., Sonderquist, D. J.,
     Wong, A. S., Symp. "Environmental Dynamics of Pesticides,"
     167th Nat. Mtg., Amer. Chem. Soc., Los Angeles, Calif.,
     Apr. 1974.
41. Nagl, H. G., Klein, W., Korte, F., Tetrahedron (1970) 26,
     5319-25.
42. Crosby, D. G., Moilanen, K. W., Arch. Environ. Contam.
     Toxicol. (1974) 2, 62-74.
43. Nagl, H. G., Korte, F., Tetrahedron (1972) 28, 5445-58.
44. Miller, L. L., Narang, R. S., Nordblom, G. D., J. Org.
     Chem. (1973) 38, 340-46.
45. Moilanen, K. W., Crosby, D. G., reported by Maugh, T. H.,
     Science (1973) 180, 578.
46. Atkins, D. H. F., Eggleton, A. E. J., in "Proc. Symp.
     Nuclear Techniques in Environmental Pollution," 521-33,
     Internat. Atomic Energy Agency, Vienna, 1971.
47. Esmen, N. A., Corn, M., Atmos. Environ. (1971) 5, 571-78.
48. Harvey, G. R., Steinhauer, W. G., Atmos. Environ. (1974)
     8, 777-82.
49. Gould, R. F., Ed., "Cleaning Our Environment - The Chemical
     Basis for Action," p. 216, Amer. Chem. Soc., Washington,
     D.C., 1969.
50. Freed, V. H., Haque, R., Schmedding, D., Chemosphere (1972),
     1, 61-66.
51. Thomas, T. C., Seiber, J. N., Bull. Environ. Contam. Toxicol.
     (1974) 12, 17-25.
52. Beyermann, K., Eckrich, E., Fresenius' Z. Anal. Chem. (1973)
     265, 4-7.
53. Bidleman, T. F., Olney, C. E., Bull. Environ. Contam.
     Toxicol. (1974) 11, 442-50.
54. Cramer, J., Atmos. Environ. (1973) 7, 241-56.

Standards Development in the Control of Hazardous Contaminants in the Occupational Environment

DOUGLAS L. SMITH and JACK E. MCCRACKEN

Office of Research and Standards Development, National Institute for
Occupational Safety and Health, Washington, D. C. 20852

Although one-third to over one-half of our waking hours are spent on-the-job, it was not until the turn of the century that there was organized concern for the worker and his working conditions.

With the advent of the space program, the population focused its attention on manned space vehicles requiring recirculative, regenerative, and detoxification procedures for prolonged, continuous (24-hour) exposures. Previously, problems encountered in closed atmospheres had been of concern almost exclusively in submarine operations; thus, the public became environment-conscious. Through the Federal regulations contained in the Clean Air Act (1), provisions have been made to control air pollution and protect the population-at-large.

In 1970, a mandate was given by Congress to assure safe and healthful working conditions for working men and women with the enactment of the Occupational Safety and Health Act (2). Specifically, the Act declares that no employee will be exposed to physical agents and substances that will cause impairment of health or functional capacities or diminished life expectancy as a result of his work experience. In addition, the National Institute for Occupational Safety and Health (NIOSH) was established and authorized by the Act to conduct research and to provide safety and health recommendations to the Department of Labor.

One of the most important provisions of the Occupational Safety and Health Act was to authorize NIOSH to develop criteria and recommended standards in the form of criteria documents. These documents specify employee health hazards for specific agents and recommend safe concentrations of airborne workplace contaminants while the employee is on-the-job. Additionally, criteria documents include recommendations for medical surveillance which may incorporate procedures for biologic monitoring; labeling and posting requirements; protective equipment to include skin, eye, and respiratory protection; educational material to inform employees of health hazards; safe work

practices to include general housekeeping and handling, storage,
and disposal recommendations; monitoring and recordkeeping
requirements; and methods for environmental sampling and chemical
analysis. To date, recommended standards have been developed for
22 chemical and physical agents listed in Table I.

Table I. NIOSH Recommended Standards Developed Under the
Occupational Safety and Health Act

Ammonia	Inorganic Lead
Arsenic	Inorganic Mercury
Asbestos	Noise
Benzene	Silica
Beryllium	Sulfuric Acid
Carbon Monoxide	Sulfur Dioxide
Chloroform	Toluene
Chromic Acid	Toluene Diisocyanate
Coke Oven Emissions	Trichloroethylene
Cotton Dust	Ultraviolet Radiation
Hot Environments	Vinyl Chloride

Certain aspects of standards development raise specific
questions of a chemical nature in evaluating an occupational
health hazard. For example, the chemical properties of the
nitrogen oxides, particularly nitric oxide and nitrogen dioxide,
are extremely important from the biologic standpoint. The toxicity of these oxides of nitrogen is generally attributed to
nitrogen dioxide, its having an irritant effect on the respiratory tract producing cough and sometimes mild dyspnea with
frequent remission of symptoms for up to 12 hours, followed by
potentially lethal pulmonary edema. Nitric oxide on the other
hand has been regarded as being nonirritant, its principal
toxicity supposedly resulting from the conversion of hemoglobin
to methemoglobin with hypoxia resulting from methemoglobin
production. Investigations of injuries from occupational exposures to these nitrogen oxides have generally discounted the
effects of nitric oxide; however, at high temperatures nitrogen
combines with oxygen to produce nitric oxide. High concentrations of nitric oxide in the working environment have been
produced by electric or gas welding operations or exhaust
from internal combustion engines. Nitric oxide may not be
oxidized to nitrogen dioxide as readily and completely as
previously believed, in view of the fact that engine exhaust is
rapidly diluted by air, thus reducing the concentration of nitric
oxide. At 10,000 ppm, 50% of nitric oxide has been calculated to
be oxidized to nitrogen dioxide in 24 seconds whereas at 10 ppm,
7 hours would be required to oxidize the same percentage (3).
Recently, nitrogen dioxide has been reported to produce
methemoglobinemia in animals, thus complicating the nitric oxide-
nitrogen dioxide picture (4). Furthermore, nitrogen dioxide

dissolves in water to form nitric acid, nitrous acid, or nitric
oxide depending upon the temperature. Because of the moist
conditions present in the lungs, it becomes extremely complicated
from a physiological standpoint as to whether toxicity is due to
nitric oxide, nitrogen dioxide, nitric acid, or nitrous acid and
a better understanding of the chemical kinetics involved is
greatly desired. Nitrous acid has been shown to have potent
effects on tobacco mosaic virus and E coli.

Biologic monitoring of urine or blood is useful as a
diagnostic practice to indicate unacceptable absorption of
materials, thus posing a possible risk of intoxication. It must
be ascertained, however, that the analytical products are due to
occupational exposures of the environmental contaminant in
question and not due to variations of metabolic constituents
resulting from normal dietary intake. For example, elevated
urinary hippuric acid levels may result from either occupational
exposure to toluene or from the ingestion of foods containing
benzoic acid or benzoates. Other valuable biologic indicators in
the urine include phenol as a measure of benzene exposure,
urinary arsenic for inorganic arsenic exposure, and both urinary
lead and delta-aminolevulinic acid (ALA) for lead exposure. In
addition to urinary determinations for possible lead exposure,
blood lead evaluations have involved measurements of delta-aminolevulinic acid dehydratase (ALAD) activities and blood lead
determinations by the dithizone method have been used for many
years. The manual dithizone method is claimed to be reliable but
requires experienced, meticulous technicians, extremely clean
equipment, and considerable time.

The use of personal protective equipment, particulary respiratory protective devices, is another important factor in the
standards setting process. The control of exposures of workers
to airborne contaminants can best be achieved through
incorporation of proper engineering controls, especially through
the use of local and general exhaust ventilation; however, some
situations occur that include special activities or nonroutine
operations for which respiratory protection is necessary. The
use of respirators in a specific hazardous environment is limited
more frequently by facepiece type and fit than by the air purification device that is employed. Dust, fume, mist, vapor, or gas
filters may have collection efficiencies as high as 15 to 20,000
ppm but quarter- and half-mask facepiece fit generally limits
respirator usage to 10 times the established environmental
standard and full facepieces are limited to 100 times the
standard. For example, the NIOSH recommended exposure standard
for benzene is 10 ppm as a time-weighted average; therefore,
half-mask usage would be limited to benzene concentrations at or
below 100 ppm and full facepiece masks to 1,000 ppm. Another
factor to consider is that no air purifying respirator should be
recommended for use in atmospheres which are judged to be
hazardous to life. Such a consideration is based on both the

toxicity and explosive properties of the material in question.
The contaminant may be toxic to the employee or the environmental
concentration may exceed the material's lower explosive limit.
In such situations the use of air-line respirators or self-con-
tained breathing apparatus is recommended. Considerable progress
has been made in recent years in the use of activated carbon for
the collection of organic vapors and gases. Metallic coatings
such as manganese or copper have been employed for the successful
sorption of gases such as chlorine, hydrogen sulfide, and
hydrogen cyanide. In addition, iodine impregnated activated
carbon is superior to nonimpregnated carbon for the isolation of
mercury vapor.

No environmental exposure standard is meaningful unless
procedures are available to collect and measure the contaminant
at concentrations which encompass the prescribed standard.
Measurements at recommended exposure levels frequently stretch
the capability of the best analytical method available. In
addition, results from an analytical method are no better than
the means by which the samples were obtained. An analytical
method having 2 to 5 percent error, when combined with a sampling
error of 10 to 15 percent, is considered excellent in the industrial situation where selectivity of sampling and specificity of
analysis are desired but often difficult to obtain. The
separation of sulfur dioxide gas from sulfuric acid mist can be
achieved rather well using a sampling train consisting of a suitable pore-size filter for particulate collection followed by gas
collection in liquid, possibly hydrogen peroxide solution.
Analysis can be performed by a sulfate titration method; however,
difficulties occur in certain industries such as smelter operations, where many soluble particulate sulfates interfere. Also,
it has been found that erroneous sulfuric acid determinations
have resulted and have been attributed to the binders used in
most glass fiber filters. The binders have reacted with the
sulfuric acid being collected on the filter.

Sampling problems contribute markedly to sampling error and
a constant effort must be exerted to keep the error to a minimum.
As much as 22 percent variation has been observed from personal
samples taken simultaneously only 8 inches apart. Liquid
bubblers are frequently necessary for sampling but are not easily
worn by workers, often hindering free movement. Battery powered
sampling pumps until recently were incapable of operating at set
flow rates over complete work shifts.

Briefly then, the development of occupational environmental
health standards for the control of hazardous contaminants
requires a critical review of existing information in conjunction
with professional evaluation of its significance to the working
situation. In addition to considering acute and chronic health
effects, related problems involve medical surveillance,
engineering controls, and sampling and analytical procedures. A
meaningful evaluation of the working environment can be achieved

only with the collection of samples that are representative of the worker's environment and with appropriate analytical chemical methods.

ABSTRACT

Under the provisions of the Occupational Safety and Health Act of 1970, NIOSH is charged to develop criteria and recommended standards for toxic materials to protect the health of workers. The standards include workplace environmental exposure limits, sampling and analytical procedures, medical surveillance, labeling and posting requirements, personal protective equipment and protective clothing, informing of employees of the hazards, work practices, and monitoring and recordkeeping procedures. In support of the evaluation of the hazardous effects of a substance, methods for sampling and analysis are an integral part of standards development, yet problems are presented in the working situation. Low environmental concentrations, physical occurrence and chemical characteristics of the contaminant, and limitations of equipment capability and laboratory reproducibility are but a few of the factors which have influenced measurements and have limited the availability of meaningful occupational exposure-effect relationships in the standard setting process. Additionally, under certain circumstances, monitoring the concentrations of absorbed substances or their metabolites (biologic monitoring) provides a valuable measurement technique to verify exposure of the individual worker. Levels determined primarily in urine or blood samples may represent unacceptable absorption of environmental contaminants posing a risk of intoxication.

Literature Cited

1. Clean Air Act of 1963, Public Law 88-206 (42 U.S.C. 1857 et seq.).
2. Occupational Safety and Health Act of 1970, Public Law 91-596 (29 U.S.C. 651 et seq.).
3. Austin, A. T.: The chemistry of the higher oxides of nitrogen as related to the manufacture, storage and administration of nitrous oxide. Br. J. Anaesth. (1967) 39, 345-50.
4. Greenbaum, R., Bay, J., Hargreaves, M. D., Kain, M. L., Kelman, G. R., Nunn, J. F., Prys-Roberts, C., Siebold, K.: Effects of higher oxides of nitrogen on the anaesthetized dog. Br. J. Anaesth. (1967) 39, 393-404.

6

Electrically Augmented Filtration of Aerosols

G. H. FIELDING, H. F. BOGARDUS, R. C. CLARK, and J. K. THOMPSON

Naval Research Laboratory, Washington, D. C. 20375

In the filtration of aerosols by fibrous media, certain electrical processes may be utilized to help effect particle capture. One such process is dielectrophoresis. Dielectrophoresis, a term coined by Pohl (1), refers to the migration of an uncharged, but polarized particle under the influence of a divergent electric field. This process is not to be confused with electrophoresis, the process which occurs in electrostatic precipitation, where a charged particle is attracted to an oppositely charged electrode. This report presents results of experiments in which dielectrophoresis was used to enhance the filtration efficiency of commercial glass fiber filter media.

An uncharged aerosol particle within a homogeneous electric field is polarized by the field, but is not subject to any displacing force due to the field. If, however, there is placed in the field a foreign body, such as a filter fiber of material whose dielectric constant is greater than one, the field becomes distorted due to polarization of the fiber. Surrounding the fiber there is a resultant field gradient with intensity increasing toward the fiber surface. An uncharged but polarized aerosol particle entering such a region of inhomogeneous field will be accelerated in the direction of increasing field intensity, i.e., toward the fiber surface, where it may be captured.

This dielectrophoretic effect described for a single fiber is multiplied many times for a fibrous filter medium placed in an electric field. Every interfiber space in the filter mat becomes a microscopic region of field inhomogeneity in which dielectrophoresis can occur. Dielectrophoresis in filtration occurs concurrently with and in addition to the usual mechanisms contributing to aerosol deposition, namely, interception, inertial impaction, and diffusion.

The first reported application of dielectrophoresis to aerosol filtration was in 1954, when the Harvard University Air Cleaning Laboratory tested a manufacturer's prototype of a proposed

68

electrified filter unit (2). Compared to that of glass fiber mats alone, the performance of this filter unit was better by factors of 1.5 to 6.5, depending upon test conditions. Apparently, such improvement was not considered to be enough to justify further work, and the device was never marketed.

In 1959, Thomas and Woodfin reported obtaining a 5-fold improvement in performance of an electrostatic precipitator by packing the void spaces between collector plates with a fibrous filter material (3). According to their paper, a major effect of this packing was to greatly decrease the free path length of a particle and thus increase the probability of its capture when deflected by the electric field.

In a 1962 paper, Rivers (4) discussed the operating principles of non-ionizing electrostatic air filters of the type patented by Dahlman in 1950 (5). Rivers derived equations for calculating the component of aerosol drift velocity due to polarization forces and compared his calculated values to typical experimental values. Agreement was best in the 2 to 5 micron particle size range. The manufacturer of the electrified filter described by Rivers claimed that aerosol penetration was decreased by a factor of 1.7 because of electrification.

The merits of different orientations of the electric field relative to the direction of air flow were discussed by Havlicek in 1961 (6). He showed both theoretically and experimentally that when the electric field is impressed parallel to the direction of air flow, the maximum electrical force on a particle reinforces the maximum aerodynamic forces leading to deposition.

In 1964, Walkenhorst and Zebel constructed an idealized model filter using many layers of nylon hosiery material with a carefully arranged array of electrodes interspersed every 10 layers (7). This filter had a low resistance to air flow because of the open structure of the knitted nylon fabric. Filtration thus depended largely upon the action of the electric field, which was oriented parallel to the direction of air flow. Walkenhorst and Zebel subsequently published extensive analyses of the performance of this geometrically regular model (8, 9, 10, 11, 12, 13).

The theoretical aspects of dielectrophoretic filtration have been extensively treated by investigations such as those mentioned. Hence, the Naval Research Laboratory effort has emphasized seeking a practical and economical application of dielectrophoresis to improve the performance of existing commercial filter media. A simple configuration has been used in which the electric field is impressed parallel to the direction of air flow.

Experimental

The filter media studied were of a type normally used for dust removal or for pre-filtration ahead of high efficiency filters. They were reinforced, nonwoven glass fiber mats 6.4 mm

thick. Three grades were used; they differed from each other in
fiber blend, packing density, and their resulting filtration capa-
bilities. The filtration characteristics of these filters, as
stated by the manufacturer (14), are shown in Table I.

Table I. Filtration characteristics of glass fiber filter media.

Filter medium	Mfr. rated flow conditions		Aerosol retention	
	Pressure drop	Air velocity	5μ dust	DOP aerosol
HP-15	0.35 in. w.g.	44 cm/sec	99 %	Not rated
HP-100	0.40 in. w.g.	20 cm/sec	99.7 %	60-65%
HP-200	0.40 in. w.g.	17 cm/sec	99.9 %	80-85%

For experimental purposes a 14 cm x 14 cm section was
mounted in a hardboard frame. This framed filter assembly was
then sandwiched between two 20-mesh stainless steel screen electrodes. An exploded view of this assembly is shown in Figure 1.

*Figure 1. Filter-electrode assembly, exploded view. Reference scale in
inches.*

The filter-electrode assembly was mounted between round-to-
square transitions in the center of a cylindrical duct 10 cm in
diameter by 250 cm in length. Air movement through the system was

provided by a canister-type vacuum cleaner at the downstream end.
The cleaner motor speed was controlled by a variable autotrans-
former. Air flow rate was measured by means of the pressure drop
across a calibrated nozzle in the duct. Provision was made also
for measuring the pressure drop across the filter.

Aerosol was introduced into a plenum at the inlet end of the
duct. Two liquid aerosols were used: (1) 0.3 micron-diameter di-
octylphthalate (DOP) generated by a vapor condensation process and
(2) nominal 1.0 micron-diameter DOP generated by an atomizer coupled to a jet impactor for removal of large drops. Aerosol concentration before and after the filter was measured with a light-
scattering photometer. Each aerosol sampling point was preceded
in the duct by a series of orifice plates for aerosol mixing.

A variable-voltage, positive-ground DC power supply was connected to the wire screen electrodes. This provided an electric
field through the filter parallel to the direction of air flow.
There was a small current, less than 0.25 microamperes, through
this circuit which was attributed to leakage through insulation.
The voltages applied were too low to generate a corona discharge
between the electrodes. Had there been a corona, the current
would have been of the order of a few milliamperes.

The experimental procedure involved the simultaneous measurement of filter penetration (or filter efficiency) and pressure
drop in the filter at a number of voltages and air flow rates.
Starting with the lowest flow rate, the unfiltered and filtered
aerosol concentrations were measured first at zero voltage and
then at successively higher voltages up to a maximum of 7 kv.
This procedure was repeated as air flow rate was increased stepwise up to the maximum appropriate for the filter.

Results

The results of the dielectrophoretic filtration study are
presented in Figures 2 through 7. Each figure shows the percentage of aerosol retained by the filter as a function of air velocity (or pressure drop) at applied voltages from 0 to 7 kv. For the
6.4 mm thickness of these filters the electric field through the
filter (in kv/cm) was 1.57 times the applied voltage.

The effect of the applied electric field in enhancing filtration efficiency has been rated numerically by means of a calculated index called the Dielectrophoretic Augmentation Factor
(DAF). This number is the ratio of the percent aerosol penetration (100% - % retention) at zero voltage to the penetration at
the voltage of interest. For example, if a filter at a given flow
rate showed a penetration of 10% at zero voltage and 1% at 7 kv,
the DAF for that set of conditions would be 10. Values of the DAF
are shown in Tables II through VII as a function of applied
voltage for each air flow rate studied.

Discussion

        Figures 2 and 3 show the aerosol retention by the HP-15 fil-
ter when challenged with 0.3 and 1.0 micron DOP, respectively.
HP-15 has a fairly open structure; hence, aerosol retention was
relatively low. At zero applied voltage the aerosol retention increased as air velocity increased. The relative increase of retention with velocity was greater for the larger aerosol than for
the smaller. This qualitatively agrees with the concept of inertial deposition increasing with velocity for these different
sized aerosol particles. Application of the electric field had
the greatest effect on efficiency at the lowest air flow rates.
The effect of the field decreased as air flow rate increased.
This is to be expected, since at higher velocities the aerosol has
less time to be influenced by the field. One might expect that
at extremely high velocities the dielectrophoretic effect would be
negligible compared to that of the inertial mechanism.
        Calculated values of the DAF are presented in Tables II and
III for Filter HP-15 challenged with 0.3 and 1.0 micron DOP, respectively. At the manufacturer's recommended flow rate of 44
cm/sec (0.35 in. pressure drop) and with 7 kv applied to the
electrodes the DAF was 2 for 0.3 micron aerosol and 4 for 1.0
micron aerosol.

Table II. Dielectrophoretic augmentation factor as a function
of voltage and air velocity in HP-15 filter medium; 0.3 micron
DOP aerosol.

Air velocity, cm/sec	Filter voltage, kv			
	2	3.5	5	7
7	2	3	6	9
14	2	2	3	5
21	1	2	3	3
33	1	2	2	3
44	1	1	2	2
56	1	1	2	2

Table III. Dielectrophoretic augmentation factor as a function
of voltage and air velocity in HP-15 filter medium; 1.0 micron
DOP aerosol.

Air velocity, cm/sec	Filter voltage, kv			
	2	3.5	5	7
7	3	6	11	23
14	2	3	5	9
21	2	3	4	7
33	1	2	4	5
44	1	2	3	4
56	1	2	2	3

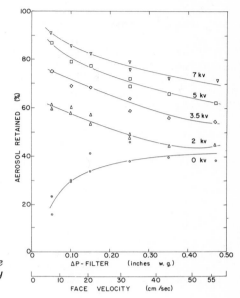

Figure 2. Influence of applied voltage upon retention of 0.3-μ DOP aerosol by HP-15 filter medium

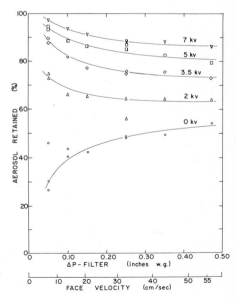

Figure 3. Influence of applied voltage upon retention of 1.0-μ DOP aerosol by HP-15 filter medium

The retention of 0.3 and 1.0 micron aerosols by Filter HP-100 is shown in Figures 4 and 5, respectively. The related DAF values are shown in Tables IV and V. Performance of this filter was

Table IV. Dielectrophoretic augmentation factor as a function of voltage and air velocity in HP-100 filter medium; 0.3 micron DOP aerosol.

Air velocity, cm/sec	Filter voltage, kv			
	2	3.5	5	7
3	8	19	95	330
5	3	13	39	120
8	3	11	28	100
13	2	6	13	42
18	2	5	9	27
26	2	4	6	14
37	2	3	4	9
46	1	2	3	6

Table V. Dielectrophoretic augmentation factor as a function of voltage and air velocity in HP-100 filter medium; 1.0 micron DOP aerosol.

Air velocity, cm/sec	Filter voltage, kv			
	2	3.5	5	7
3	30	110	300	1100
5	6	30	95	360
8	4	18	50	170
13	3	10	20	50
18	2	6	13	35
26	2	4	8	18
37	2	3	5	11
46	1	2	3	7

qualitatively similar to that of HP-15, but the efficiency was higher throughout. Again, the dielectrophoretic effect was greatest at the low flow rates and decreased as velocity increased. From Tables IV and V one can interpolate a value of the DAF for the manufacturer's recommended flow rate of 20 cm/sec (0.40 in. pressure drop). At this flow rate and with an applied voltage of 7 kv the DAF is 21 for 0.3 micron aerosol and 28 for 1.0 micron aerosol.

The retention of 0.3 and 1.0 micron aerosols by Filter HP-200 is shown in Figures 6 and 7, respectively. HP-200 was the most efficient filter of the three tested. Still, the augmentation effect of the electric field was quite significant. At the manufacturer's recommended flow rate of 17 cm/sec (0.40 in. pressure drop) and with an applied voltage of 7 kv the DAF interpolated from Tables VI and VII is 19 for 0.3 micron aerosol and 30 for 1.0 micron aersol.

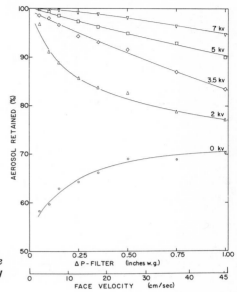

Figure 4. Influence of applied voltage upon retention of 0.3-µ DOP aerosol by HP-100 filter medium

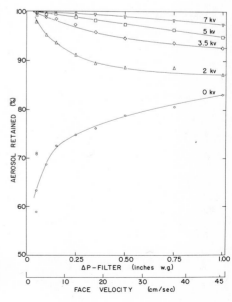

Figure 5. Influence of applied voltage upon retention of 1.0-µ DOP aerosol by HP-100 filter medium

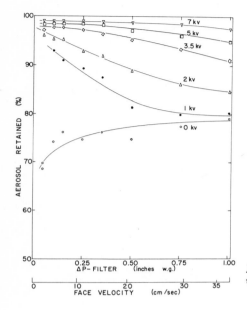

Figure 6. Influence of applied voltage upon retention of 0.3-μ DOP aerosol by HP-200 filter medium

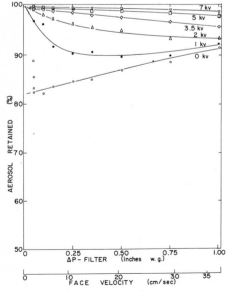

Figure 7. Influence of applied voltage upon retention of 1.0-μ DOP aerosol by HP-200 filter medium

Table VI. Dielectrophoretic augmentation factor as a function of voltage and air velocity in HP-200 filter medium; 0.3 micron DOP aerosol.

Air velocity, cm/sec	Filter voltage, kv				
	1	2	3.5	5	7
2	5	8	11	18	28
4	4	5	12	15	22
6	3	5	10	15	22
11	2	4	9	16	24
15	2	3	7	13	20
21	1	2	5	10	18
29	1	2	3	6	13
37	1	1	2	4	8

Table VII. Dielectrophoretic augmentation factor as a function of voltage and air velocity in HP-200 filter medium; 1.0 micron DOP aerosol.

Air velocity cm/sec	Filter voltage, kv				
	1	2	3.5	5	7
2	6	26	66	127	380
4	5	11	34	65	138
6	2	7	21	43	50
11	2	5	12	23	37
15	2	4	10	19	43
21	1	3	6	13	26
29	1	2	4	7	18
37	1	1	2	4	8

The relative improvement in filtration performance due to the applied electric field was notably greater for the HP-100 filter than for the HP-15. The HP-100 blend contains more fine fibers than the HP-15 and is more densely constructed. This would have offered more opportunity for dielectrophoresis to occur, all other things being equal. In terms of the DAF, the relative improvement shown by Filter HP-100 was about the same as that shown by Filter HP-200 at their respective rated flow conditions. Because of differences in initial filtration efficiency, the relative improvement caused by dielectrophoresis was greater at low flow rates for Filter HP-100 than for Filter HP-200.

The possible effect of charged aerosol particles has not been addressed in this study. If any aerosol particles had been inadvertently charged, the filtration improvement attributed here entirely to dielectrophoresis would have been due in part to coulombic forces.

Conclusions

Application of an electric field through a glass fiber dust

filter medium effects a substantial improvement in filtration efficiency by means of dielectrophoresis. Further work is required
to learn the most effective conditions of air flow rate, fiber
blend, density, and applied voltage. Further investigation is
also needed to test the applicability of dielectrophoresis to
filter media other than glass fiber and to aerosols other than
DOP.

Literature Cited

1.   Pohl, H. A., J. Appl. Phys., (1951), 22, 869-871.
2.   Billings, C. E., Dennis, R., and Silverman, L., "Performance
      of the Model K Electro-Polar Filter," Air Cleaning
      Laboratory, Harvard University School of Public Health,
      Boston, Massachusetts, Report NYO-1592, 1954.
3.   Thomas, J. W., and Woodfin, E. J., A.I.E.E. Tr., Pt. II,
      (1959), 78, 276-278.
4.   Rivers, R. D., A.S.H.R.A.E.J., (1962), 37-40.
5.   Dahlman, V., "Electrical Gas Cleaner Unit," U. S. Patent No.
      2,502,560, 1950.
6.   Havlicek, V., Int. J. Air and Water Poll., (1961), 4,
      225-236.
7.   Walkenhorst, W., and Zebel, G., Staub, (1964), 24, 444-448.
8   Zebel, G., J. Colloid Sci., (1965), 20, 522-543.
9.   Zebel, G., Staub, (1966), 26, 18-22.
10.   Zebel, G., Staub, (1969), 29, 21-27.
11.   Walkenhorst, W., Aerosol Science, (1970), 1, 225-242.
12.   Walkenhorst, W., Staub, (1969), 29, 1-13.
13.   Davies, C. N., Filtration and Separation, (1970), 692-694.
14.   Farr Company, "HP Air Filters," Technical Data Bulletin
      B-1300-4K, Los Angeles, California, 1969.

Experimental and Theoretical Aspects of Cigarette Smoke Filtration

CHARLES H. KEITH

Celanese Fibers Co., Box 1414, Charlotte, N. C. 28232

A cigarette filter is a seemingly simple filtration device which has come to be a major component in the tobacco industry in the past twenty-five years. Like many filter systems, it utilizes well established fiber filtration mechanisms to remove particles from the smoke stream, and it has an adsorbative capacity for some of the vaporous smoke components. However, its small and defined size - a cylinder of about 8mm diameter and 15 to 30mm length, and its extensive use as a consumer item - about five hundred billion filter cigarettes are consumed in the U. S. each year - impose special requirements. The most obvious of these requirements is that a cigarette filter must be a partial filter, as very few if any cigarette smokers are interested in brands with no appreciable taste or impact. At the present time, most popular brands of cigarettes are equipped with filters which remove 40 to 50% of the smoke stream. A further consumer acceptance requirement is that the draw resistance or pressure drop of the filter cigarette be low, as brands with total pressure drops in excess of one hundred and fifty millimeters of water have been found to be unacceptable by the public. The filter pressure drop should thus not exceed one hundred millimeters of water at a standard flow of 17.5 ml/sec. Consumer preference also dictates that the filter should have a smooth, uniform appearance and this as well as manufacturing constraints make it necessary that the filter be reasonably firm, both initially and throughout the smoking process.

Because of the large numbers of filters manufactured each year, the filter rod-making process and the combining process must be done at speeds up to three to four thousand units per minute. This requires that the filter material be taken from a bale, fluffed or opened, and made into cylindrical paper-wrapped rods at speeds of up to four hundred meters per minute. The material thus has to be reasonably strong, uniform, and readily handeable. These stringent requirements severely limit the choice of cigarette filter materials. The net result is that the vast majority of filters in this country and increasingly throughout the world consist of a tow or bundle of crimped cellulose acetate continuous filaments.

Before discussing the properties of this and some other types of filters, it is desirable to consider whether a filter is effective from a health standpoint, since this question is always present. The filter does remove substantial quantities of the complex mixture of components in tobacco smoke, commonly called tar, and it removes considerable amounts of physiologically active components such as nicotine and phenol. Epidemiological studies by Bross (1) and Wynder (2) have shown that filter cigarette smokers run a significantly lesser chance (50 to 60%) of contracting some respiratory diseases associated with smoking. Dontenwill (3) has recently shown that acetate filters provide a significant degree of protection in inhalation experiments in hamsters. Thus it is apparent that a filter does reduce the amount of smoke ingested, and it is effective to a degree in the context of public health.

As indicated previously, the usual cigarette filter consists of a cylindrical bundle of crimped acetate fibers as illustrated in Figure 1. In general, there are some ten to fifteen thousand filaments present in the bundle which extend through the filter parallel to the axis of the cylinder and direction of smoke flow. The fibers have ten to twenty zig-zag crimps per inch, so that on the average the fibers are oriented at a thirty- to forty-degree angle to the flow direction. The packing of crimped fibers is such that the volume fraction occupied by fibers is about .1, and there are a number of contact points between fibers. Under normal smoking conditions which consist of 35ml puffs of two seconds duration taken once a minute, the flow is laminar with a Reynolds number of less than 1.

With such a laminar flow regime, it is possible to theoretically predict the pressure drop and particle removal efficiency using an idealized filter model such as that diagrammed in the previous figure. In the case of pressure drop, the Happel (4) solutions of the Navier-Stokes equations for flow parallel and perpendicular to the fiber can be effectively combined to provide an overall pressure drop equation for fibers oriented at any angle to the flow direction. This model assumes that the pressure drop is generated by drag of the moving gas stream on the stationary fibers, and that the fibers have a regular orientation at the average crimp angle. It allows for the noncircular cross-section of the fibers and for fiber touching by means of corrections determined from surface area data and microscopic counts of fiber bundles. The complete pressure drop equation is given in Figure 2 with a description of the variables included. Figure 3 presents a comparison of observed and calculated pressure drops for a wide variety of filter rods, which are six filters long. The agreement is excellent as the points are closely clustered about the 1:1 line. The standard deviation is 23mm or 6.5% for a mid-range rod of 350mm pressure drop.

From use of the pressure drop equation, it is found that pressure drop increases linearly with flow rate, filter length,

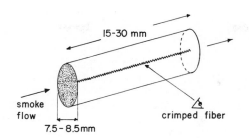

Figure 1.   Model of a cigarette filter

$$\Delta p \;=\; \frac{7.34\times10^{4} \cdot \pi \cdot \rho \cdot \mu \cdot Q \cdot L \cdot S_x}{F \cdot \delta \cdot b \cdot S_o} \cdot \frac{\alpha}{(2\alpha^{2}-1-\ln\alpha) + (4\alpha-3\alpha^{2}-2-\ln\alpha)\cdot\cos\Theta}$$

Where: $\alpha = \dfrac{W}{F \cdot L \cdot \rho}$; $S_o = \dfrac{3.36\times10^{3}}{\delta^{\frac{1}{2}} \rho^{\frac{1}{2}}}$; $\cos\Theta = \dfrac{T \cdot L}{9\times10^{5} \cdot W}$

Δp	=	Pressure Drop (mm H_2O @ 17.5 cm^3/sec. flow
ρ	=	Density of the fiber polymer (g/cm^3)
μ	=	Viscosity of the fluid (poise)
Q	=	Volumetric flow rate (cm^3/sec)
L	=	Filter length (cm)
S_x	=	Specific surface area of the fiber (cm^2/g)
α	=	Volume fraction occupied by fiber (unitless)
F	=	Cross-section area of the filter (cm^2)
δ	=	Fiber denier per filament (g/9x10^5 cm)
b	=	Agglomeration factor (unitless) = no. of filaments/no. of filament bundles
S_o	=	Specific surface area of equivalent cylindrical fiber (cm^2/g)
Θ	=	Average crimp angle (degrees)
W	=	Fiber weight (g)
T	=	Total denier of the filter = δ x no. of filaments (g/9/x10^{5}cm)

Figure 2.   Pressure drop equation

and fiber surface area. Increasing fiber packing or weight gives
a more than linear increase in pressure drop and is the most important variable in determining filter pressure drop. Increments
in cross-sectional filter area, non-uniformity of fiber distribution and fiber denier decrease pressure drop in a hyperbolic
fashion with the first two variables having a greater effect than
fiber denier. Increasing crimp angle increases pressure drop and
has less effect than the other variables. All of these findings
are consistent with experimental observations.

Turning now to filtration theory, the size of smoke particles
and their neutral or lightly charged character reduce the number
of important filtration mechanisms to three. These are illustrated in Figure 4 and consists simply of diffusional capture,
direct interception of the 0.1 to 1 micron particles by the 20 to
30 micron fibers, and to a lesser extent inertial impaction. The
technique that we have used to compute particle removal efficiencies is to compute the fraction of particles of a given size which
will be intercepted by or touch a single fiber by the combination
of the three filtration mechanisms. These are then integrated
over the range of particle sizes in smoke to determine the single
fiber filtration efficiency. The final step is to suitably sum
the single fiber efficiencies over all the fibers along the length
of the filter to obtain an overall particle removal efficiency.
The equation for the single fiber efficiency, derived from the
Happel solutions for a particle approaching the fiber at right
angles is given in Figure 5. To utilize this equation it is necessary to determine two variables, the diffusion radius and the
effective fiber radius. The latter can be approximately determined several ways. The simplest but least satisfactory method is
to assume that the fibers are cylinders, which is usually not the
case in cigarette filters, and directly compute the radius from
the fiber denier and density. A better approximation is obtained
by considering that the fiber has a regular, non-circular geometric cross-section such as a Y, X, or I beam. The effective
radius is then considered to be the radius of the circle circumscribed around the geometric shape, and may again be readily computed. The best method utilizes pressure drop data and the
pressure drop equation previously given. Solving this equation
numerically for the volume fraction occupied by fiber leads to an
effective radius by division by the number of filaments present.
This technique has an advantage in that it compensates for non-
uniformities in fiber cross-section, fiber distribution, and
touching fibers. As shown in Figure 6, the effective or filtration radius can be considered as a circumscribed circle around the
fiber. Particles reaching this circle are considered to be ir-
reversibly collected either by contacting one of the protruding
fiber lobes or by entering and remaining in the stagnant air
spaces spaces between the lobes. This effect was clearly demonstrated recently in the work of Morie, Sloan and Peck (5) who
showed that the primary deposition in normal cigarette filters is

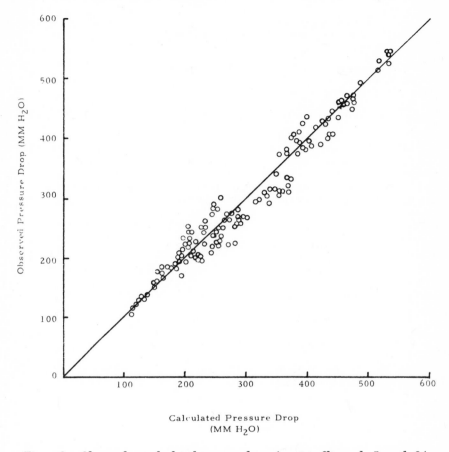

Figure 3.   Observed vs. *calculated pressure drop of acetate filter rods. Length, 84–120 mm; circumference, 23.8–27.2 mm; denier, 3–8.*

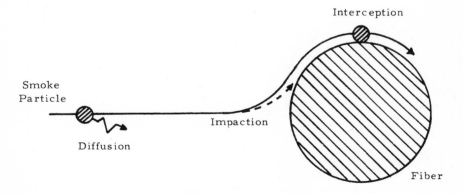

Figure 4.   Representation of filtration mechanisms

$$e_f = \frac{(1 + I) (r_p + r_d)^2}{a^2} \cdot \frac{\alpha^{\frac{1}{2}}}{(1 + \frac{1}{2} \frac{(1 + \alpha)}{(1 - \alpha)} \cdot \ln\alpha)}$$

Where: $\quad$ I $\quad$ = $\quad$ An inertial parameter = $2/9 \cdot \dfrac{C \cdot \rho \cdot r_p^2 \cdot Q}{\mu \cdot a \cdot F \cdot (1 - \alpha)} \cdot (.22 + 4\alpha)$

$\quad e_f \quad$ = $\quad$ Single fiber filtration efficiency (unitless)

$\quad r_p \quad$ = $\quad$ Smoke particle radius (microns)

$\quad r_d \quad$ = $\quad$ Radial distance that the particle diffuses during its travel around the fiber (microns)

$\quad \alpha \quad$ = $\quad$ Volume fraction occupied by fiber (unitless)

$\quad$ a $\quad$ = $\quad$ Effective fiber radius (microns)

$\quad$ c $\quad$ = $\quad$ Cunningham slip flow correction (unitless)

$\quad \rho \quad$ = $\quad$ Density of the particle (g/cm^3)

$\quad$ Q $\quad$ = $\quad$ Volumetric flow rate (cm^3/sec)

$\quad$ F $\quad$ = $\quad$ Cross-section area of the filter (cm^2)

$\quad \mu \quad$ = $\quad$ Viscosity of the fluid (poise)

Figure 5. Single fiber efficiency equation

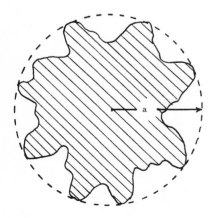

Figure 6. Effective radius of a fiber

at the tips of the fiber lobes, but in special filters with
fibers oriented parallel to the smoke stream, the deposition was
mainly in the valleys between the lobes.

Knowing the effective radius, it is possible to approximately
calculate the time that it takes for a particle to travel around
the fiber. Using Langmuir's (6) equation for the average displacement of a particle undergoing Brownian motion, the diffusion
radius can be calculated and utilized to further compute single
and overall particle removal efficiencies.

To demonstrate the agreement between theoretically derived
particle removal efficiencies and experimentally measured filtration efficiencies, the nicotine removal efficiency (NRE) of a wide
range of filters was measured and compared with the computed
efficiencies as shown in Figure 7. Nicotine was chosen for this
analysis because it is a readily measurable component of tobacco
smoke and in normal cigarettes about 90% of it is found in the
particulate phase.

In Figure 7, the plotted points fall somewhat below and to
the right of the 1:1 correlation line, and have a regression line
as indicated. There are two reasons for this departure. The
first is that not all of the nicotine is present in the particles
and that this vapor phase component can be adsorbed by the acetate
filter material. Lipp (7) has found that acetate filters remove
some 70% of vapor phase nicotine, so that the 10% of the nicotine
in the vapor phase would be more thoroughly removed for most of
the filters tested, thereby increasing the overall NRE values. At
high particle removal efficiencies this effect becomes less important and closer agreement is observed. The second reason is
that the particle removal efficiencies were calculated using fibers perpendicularly oriented to the smoke stream. Although not
tractable mathematically, fibers oriented at other angles would be
expected to increase the diffusional filtration by increasing the
travel time around the fiber without greatly affecting the other
filtration mechanisms. Thus it would be expected that the particle removal efficiencies should be somewhat higher. In this regard, it is found from the equations that diffusional capture is
the most important filtration mechanism, accounting for about 65%
to 68% of the total filtration. Direct interception accounts for
30 to 35% and inertial impaction is relatively unimportant, contributing only 1 to 1.5% to the overall filtration process.

There are several measures of overall filter performance besides nicotine removal efficiency and pressure drop. One of these
is smoke removal efficiency (SRE), which is defined as the amount
of weighable smoke captured by the filter divided by the sum of
material collected by the filter and an efficient, standard collection trap placed between the filter cigarette and the smoking
machine. Another is a tar removal efficiency (TRE), which is
similarly computed from the collected and delivered weights of a
tarry residue defined in this country as being the weight of col-
lectable smoke less the analyzed weights of water and nicotine.

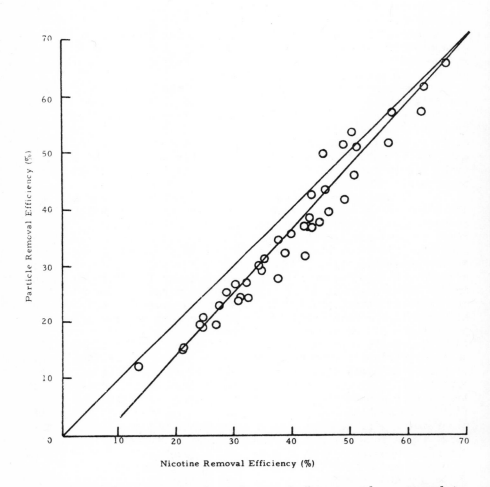

Figure 7.   Correlation of computed particle removal efficiencies with experimental nicotine removal efficiencies

Each of these efficiencies is different because of the complex physical and chemical nature of smoke and its interaction with filter media. For example, smoke contains 15 to 25% water which is distributed between the particulate and vapor phases. Since most filtering materials readily adsorb water, their SRE values which include water are considerably higher than their nicotine or tar removal efficiencies. This is illustrated in Figure 8 which shows the effect of fiber size on removal efficiency over a practical range of filter pressure drops. For the smaller fiber, 2 dpf filter SRE changes from 49 to 59% over a 40 to 100mm pressure drop range. TRE, which excludes water but includes other volatile and semi-volatile smoke components goes from 42 to 53%, while NRE which is essentially a measure of mechanical efficiency goes from 34 to 48%. Similar changes at lower efficiencies are observed for a coarser five denier per filament filter. A further demonstration of this effect and illustration of the considerable effect of filter length on performance is given in Figure 9. Here doubling the filter length at any given pressure drop increases the SRE by 15 to 19 units. The effect on NRE is less pronounced because of the water adsorption effect on SRE but is still substantial at 10 to 13 units. Other filter variables besides filter length, pressure drop and fiber size have relatively little effect on removal efficiency. Variables such as circumference, fiber cross-section, and fiber orientation and packing, all strongly affect pressure drop, but when this is held constant, there is little or no effect on removal efficiency.

With a general understanding of the mechanical filtration process in conventional filters, some of the properties of other types of cigarette filters can be briefly discussed. Another fairly common type of cigarette filter consists of a fluted or corrugated sheet of cellulose paper which is bundled into a cylindrical rod with the corrugations along the axis of the filter. This type of filter operates mechanically very much like a conventional filter, but the folded corrugations provide channels which reduce the overall pressure drop of the filter to reasonable levels without seriously deteriorating its particle removal efficiency. To be effective, this type of filter must have an essential equivalence between the pressure drop along a channel and that between channels. Without such a balance in pressure drops, it would either be like a random array of fibers with a high pressure drop or like a series of capillaries with relatively impermable walls which would have a low pressure drop and a very low removal efficiency. Another type of filter employs an intricate construction to channel the smoke across dense fiber mats with face areas larger than the cross-section of the cylindrical cigarette. Others utilize ventilation either before, in, or after, a fibrous filter plug to dilute the smoke stream and thereby reduce the tar delivery of the cigarette. A final type of filter utilizes granular or powdered adsorbents such as charcoal in conjunction with fibers or in separate chambers to adsorb gaseous and

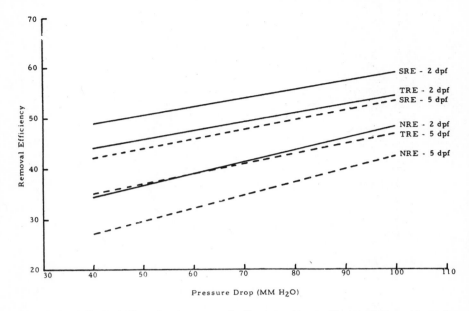

Figure 8.   Effect of fiber size on removal efficiency; 20 mm filters—24.8 mm circumference; 2 and 5 denier per filament

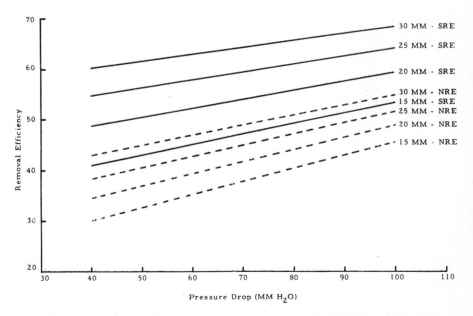

Figure 9.   Effect of filter length on removal efficiency—2.0 dpf fiber, 24.8 mm circumference

vaporous components from the smoke stream. Except for this last type of filter, the filtration properties and mechanisms are similar to those of conventional acetate filters.

In the preceeding discussion, the filtration properties for removing aerosol particles from the smoke stream have been outlined from a theoretical and experimental viewpoint. As has been mentioned, filters also remove condensible gases and vapors from the smoke stream by physical or chemical adsorption. It is this property which gives rise to selective filtration of some components of the smoke mixture. If, as all evidence available indicates, the filtration of particles is a non-reversible process in which the captured particles are not reintroduced into the smoke stream, then the greater or lesser removal of some smoke components must depend on vapor transfer processes, i.e. condensation, distillation, and physical and chemical adsorption. That this does occur has already been shown in the comparison of removal efficiencies, where the elevation of SRE and TRE values above the NRE values indicates a selective removal of water and volatile components of the tar through adsorption on the filtering material. In fibrous filters, it has been found that there is a considerable potential for selective filtration of materials with boiling points between about 100 and 300°C which are soluble or partially soluble in the filtering material. As examples of this type of selective filtration it has been found that acetate filters remove 70 to 80% of phenol from cigarette smoke, 60 to 75% of water, and 70 to 75% of cresols from the smoke stream while only removing about 35 to 40% of the particulate matter(8). This capacity for selective removal of high boiling, soluble components can be enhanced or retarded by the addition of plasticizers and reactive chemicals to the filter. For example the removal of phenol by acetate filters can be increased to about 90% by the addition of glycerol triacetate and polymeric glycols to the surface of the filter material. In the case of nicotine, the addition of bases to the filter material decreases its removal efficiency from about 35% to 25% or less. This occurs because the basic additive reacts with the nicotine salts in the captured smoke particles liberating free nicotine which distills back into the smoke stream.

As indicated previously, some types of filters contain added adsorbents such as charcoal. These types of filters are designed to remove another range of volatile smoke components, generally consisting of condensible vapors of materials with boiling points between 0 and 100°C. A number of irritating and potentially harmful smoke components such as acetaldehyde, acrolein, hydrogen cyanide, and formaldehyde are thus selectively removed by this type of adsorbent filter. These components, which are largely unaffected by conventional filters, can be removed with efficiencies of 60 to 95% depending on the component, the amount and kind of adsorbent, and the presence of surface treatments on the adsorbent. By combination of adsorbents with fibrous filters a

broad range of materials can be selectively removed from the
smoke stream(8).

In conclusion, we find that a cigarette filter, which is
probably the largest volume use of filtering material in industry,
is an effective means of modifying tobacco smoke. It removes
about one half of the smoke stream by a theoretically tractable
combination of filtration mechanisms. In addition, conventional
and specially fabricated filters show a degree of selectivity for
various volatile smoke components.

Literature Cited

1. Bross, I.D.J., Nat. Cancer Inst. Monograph (1968) 28, pp. 35-40.

2. Wynder, E.L., Mabuchi, K. and Beattie Jr.,E.J. J.A.M.A. (1970)
    213, pp. 2221-2228.

3. Dontenwill,W., Chevalier,H.J., Harke,H.P., Lafrenz, U.,
    Reckzeh,G., and Schneider,B., J. Nat. Cancer Inst. (1973) 51
    pp. 1781-1807.

4. Happel,J., A. I. Ch. E. Jour. (1959) 5, pp. 174-177.

5. Morie,G.P., Sloan,C.H. and Peck,V.G., Beitr. Tabakforsch(1972)
    7, 99-104.

6. Langmuir,I., OSRD Report 865 (1942), cf. Chen,C.Y., Chem.Rev.
    (1955) 55, pp. 595-623.

7. Lipp,G., Beitr. Tabakforsch (1965) 3, pp. 109-127.

8. George,W. and Keith,C.H., "Tobacco and Tobacco Smoke" pp.577-
    622, Academic Press, New York, 1967; cf. Keith, C.H., "The
    Chemistry of Tobacco Smoke", pp. 149-166, Plenum Press,
    New York, 1972.

Aerosol Filtration by Fibrous Filter Mats

WILLIAM S. MAGEE, JR. and LEONARD A. JONAS
Edgewood Arsenal, Aberdeen Proving Ground, Md. 21010
WENDELL L. ANDERSON
Naval Weapons Laboratory, Dahlgren, Va. 22448

Abstract
        A semi-empirical formalism for analyzing aerosol filtration
by fibrous filters is modified and compared to several theories.
Although approximate, the formalism introduced by Dorman [3,4,5]
proves to be useful and instructive. The formalism is applied
to analysis of penetration-velocity profiles for ten different
types of filters, each challenged by dioctyl phthalate (DOP)
aerosols of four distinct droplet sizes (0.26, 0.28, 0.30,
0.32 μm) in the velocity range, 7.2 to 141.0 cm/sec.
Quantitative comparison is made with a simple theoretical model
of Fuchs[6] Reasons for discrepancies are considered in terms of
modern concepts of filtration.

Introduction

        Even with the alternatives offered by current sophisticated
technology fibrous filters are the choice for the efficient
removal of submicron sized aerosol constituents from air streams.
Although simple in design and application such filters involve
removal mechanisms which result in complicated mathematical
models for even the simplest idealized conditions. Consequently
both the analysis of filtration data and the quantitative
rationale for comparison of filters are often ad hoc in nature
and limited in applicability. Moreover the better theoretical
models, although reproducing some parametric dependencies, often
give results differing from experimental data by orders of
magnitude.
        The present study was initiated to provide at least a semi-
empirical formalism for the study and assessment of aerosol
filtration by fibrous filters. This paper describes the basis
of the formalism and its application to filtration data for ten
different types of fibrous filters. The filtration data consists
of penetration-velocity profiles for each of the types of filters
challenged by dioctyl phthalate (DOP) aerosols with each of four
distinct droplet sizes (0.26, 0.28, 0.30, 0.32 μm). Profiles

cover the linear face velocity range, 7.2 to 141.0 cm/sec. The
formalism characterizes the data in terms of three removal
mechanisms: inertial impaction, diffusion and interception. A
comparison is made to theoretical results given by a simplistic
model by Fuchs. Reasons for discrepancies are discussed.

Theory

As a starting point we adopt a modification [1,2] of a semi-
empirical treatment by Dorman [3,4,5] which states that the percent
penetration, P, of the filter is given by the relation

$$\log P = 2 - k_I L V^x - k_D L V^{-y} - k_R L \qquad (1)$$

where L is the thickness of the filter, V is the linear face
velocity, k_I, k_D and k_R are Dorman parameters characterizing the
mechanisms of inertial impaction, diffusion and interception
respectively, and x and y are numbers chosen either from theory
or from fitting equation (1) to data. Part of the present study
is to rationalize choices for x and y, various authors [2] having
found $1 \leq x \leq 2$ and $1/2 \leq y \leq 2/3$.

In the original Dorman procedure the velocity, V_o, for which
P is a maximum is determined by inspection of graphs for $\log P$ vs
V. Often data for filters present either broad plateaus with no
distinct maximum or no maximum occurs. Such data cannot be
analyzed by the original Dorman procedure. Our modification [2]
eliminates this deficiency and consists of using the identity
transformation

$$V^x = (V^{-y})^{-x/y} \qquad (2)$$

in equation (1), of setting the derivative with respect to V^{-y} of
the transformed equation equal to zero and of solving for Vo.
These manipulations yield the following relationships:

$$\frac{d \log P}{d V^{-y}} = \frac{x}{y} L k_I \left[V^{x+y} - V_o^{x+y} \right] \qquad (3)$$

and

$$k_D = \frac{x}{y} k_I V_o^{x+y} \qquad (4)$$

From linear regressions of $\frac{d \log P}{d V^{-y}}$ on V^{x+y}, k_I is obtained from
the slope and V_o from the intercept according to equation (3).
This systematically determines V_o, k_I, k_D from equation (4) and
k_R from equation (1), thus characterizing the filter.

In actual practice the derivative $\frac{d \log P}{d V^{-y}}$ must be estimated
from discrete penetration-velocity data sets. Any numerical

differentiation technique can be used as a means of estimation. For example, Lagrangian three-point or five-point interpolation techniques can be used. However, for ease of calculation we used a simple averaging technique in the form

$$\frac{\Delta \log P}{\Delta V^{-y}} = \frac{x}{y} Lk_I \left[V_m^{x+y} - V_o^{x+y} \right] \tag{5}$$

where V_m is defined by the relation

$$V_m^{x+y} = \left[\frac{V_i^{-y} + V_{i+1}^{-y}}{2} \right]^{-\left(\frac{x+y}{y}\right)} \tag{6}$$

In the analysis of data these equations can be used in computer algorithms to obtain "best fits" by systematically varying the parameters. However, we have chosen to rely upon theory to give values for x and y and then calculate k_I, k_R and k_D. The simplest model we have used is that introduced by Fuchs.[6] Neglecting all mechanisms except inertial impaction, diffusion and interception we may write Fuchs' equations for assumed Poiseulle flow among the fibers in the form

$$\log P = 2 - 0.4343 \left(\alpha_R + \alpha_D + \alpha_I \right) \tag{7}$$

where the dimensionless coefficients α_I, α_D, α_R are given by the relations

$$\alpha_I = \frac{\tau VL}{2(3)^{\frac{1}{2}}h^2} \tag{8}$$

$$\alpha_D = \frac{2D^{2/3}_L V^{-2/3}}{3^{1/6} (d_f/2)h^{2/3} (1+\frac{2h}{d_f})^{5/3}} \tag{9}$$

and

$$\alpha_R = \frac{3 (d_p/2)^2 L}{2(3)^{\frac{1}{2}} h^2 (\frac{d_f + 2h}{2})} \tag{10}$$

In these equations τ is the relaxation time determined by viscous media forces on a particle, h is one half the average distance between nearest neighbor filtering fibers, D is the particle diffusion coefficient, d_f is the diameter of a filtering fiber and d_p is the diameter of an aerosol particle.

Comparison of these relations with those generalized from Dorman leads to the following identifications for k_I, k_D and k_R =

$$k_I = \frac{0.1254 \ \tau}{h} \tag{11}$$

$$k_D = \frac{1.44663 \ D^{2/3}}{h^{2/3} d_f^{2/3} (d_f + 2h)^{5/3}} \tag{12}$$

and

$$k_R = \frac{0.1880 \ d_p^2}{h^2 (d_f + 2h)} \tag{13}$$

Using Fuchs' early model of a fibrous filter having parallel cylinders (fibers) positioned in a checkered arrangement at a

distance $2h$ from each other, we can show that $2h$ is simply
related to the ratio of the void (ε) to fiber (σ) fractions of
the filter in the form

$$2h = df \frac{\varepsilon}{\sigma} \tag{14}$$

This is calculated from consideration of the following:
The volume fiber fraction $\sigma = 1 - \varepsilon$ (15)
The volume void fraction is simply related to the bulk density
ρ_{FM} of the filter mat and the density ρ_f of the fiber material by

$$\varepsilon = 1 - \frac{\rho_{FM}}{\rho_f} = 1 - \sigma \tag{16}$$

Under assumption that the fiber array is isotropic, the volume
and area fiber fractions are equivalent. This leads to the form

$$\varepsilon = \frac{2h}{2h + df} \tag{17}$$

from which (14) follows through (15).

The relaxation time for a spherical particle of mass m is
given by Fuchs as

$$\tau = \frac{m}{3\pi dp\, \eta} \tag{18}$$

where η is the viscosity of air. Relating the mass m of the
particle to its density ρ_p and volume converts (18) to

$$\tau = \frac{d_p^2 \rho_p}{18\eta} \tag{19}$$

For aerosol particles with diameters comparable to the mean
free path of the suspending gas molecules the diffusion coefficient D is given [7] by

$$D = \frac{RT\,(1 + 2A\ell/dp)}{No 3\pi_\eta dp} \tag{20}$$

where R is the gas constant, T is the absolute temperature, ℓ is
the mean free path of air "molecules", No is Avogadro's number
and A is a numerical factor equal to unity. The factor in
parentheses can be written as

$$1 + \frac{2A\ell}{dp} = 1 + \frac{2x10^{-4}}{P_a dp}\left[6.32 + 2.01\exp(-1095 P_a dp)\right] \tag{21}$$

where P_a is the atmospheric pressure in cmHg.

This completes the set of equations used in our analysis.
First, with $x = 1$ and $y = 2/3$ equation (5) is used with (6) to
obtain Vo and k_I . k_D is then obtained via (4) and k_R from (1).
This characterizes the system in terms of the penetration data.

Independently for comparison equations (11), (12) and (13)
(together with (14), (16), (19), (20) and (21)) are used to
calculate k_R, k_D and k_I from the physical parameters of the
filters and aerosols. This characterizes the system in terms of
physical properties.

Materials

The ten filter mats studied represented a wide range of fiber types and compositions. Filter mat Type 5 was developed at Edgewood Arsenal in the 1940's and is a mixture of coarse, supporting matrix fibers and Blue Bolivian crocidolite asbestos as the aerosol filtering fiber. The N11, N13 and N15 mats were developed by Naval Research Laboratory, and each contained an 84.1 to 15.9 mixture of viscose and B.B. asbestos. They differ in the degree of acid beating of the asbestos fibers. The remaining six filter mats were specially fabricated by Naval Research Laboratory for studies of aerosol filtration and were composed entirely of their fiber designations. The fiber compositions used for the formation of the filter mats are shown in Table I. From weight per unit area, mat thickness and fiber densities, physical properties were calculated and are shown in Table II. Using these and average values for the diameters of the filtering fibers in (14) and (16), basic geometric parameters were determined and are shown in Table III. Pertinent properties of the DOP aerosol were calculated from (19), (20) and (21) and are shown in Table IV.

Equipment and Procedures

The aerosol test apparatus, the DOP aerosol and procedures are detailed in our earlier paper [2]. All experimental tests were performed at Naval Research Laboratory, Washington, D.C.

The conditions of the tests were the following: temperature, $25^{\circ}C$; pressure, 76 cmHg, $N_0 = 6.023 \times 10^{23}$ molecules mole^{-1}, $R = 8.3143 \times 10^{7}$ ergs mole^{-1}deg^{-1}, $\eta = 1.818 \times 10^{-4}$gcm^{-1}sec^{-1}, $\ell = 6.53 \times 10^{-3}$ cm, and $\rho_p = 0.986$g cm^{-3}.

Results and Discussion

The aerosol penetration data for ten fibrous filters, obtained by measurement of the percent penetration of four discrete diameters of DOP (0.26, 0.28, 0.30 and 0.32 μm) in the linear face velocity range 7.2 to 141 cm/sec are given in Table V. Our modified Dorman procedure embodied in equations (1), (4), (5) and (6) when applied to these data results in the Dorman parameters k_I, k_D and k_R shown in Table VI. Values of $x = 1$ and $y = 2/3$ were used for direct comparison with the Fuchs' model and gave excellent correlation coefficients for the procedural linear regression.

Dorman parameters calculated from physical and geometric properties of the aerosol and mats through equations (11), (12) and (13) are shown in Table VII. Surprisingly in spite of the simplicity of the Fuchs' model, the experimental and calculated Dorman parameters are often the same order of magnitude. The

TABLE I. COMPOSITION OF FIBROUS FILTER MATS

Fiber Type	Fiber diam. cm x10^4	% Composition of filter mats									
		Type 5	N11	N13	N15	Esparto	Visc. 1.5D	Visc. 3.0D	A	AA	AAA
Cotton floc	17.0	58.2									
Viscose rayon	12.0–17.0	34.0	84.1	84.1	84.1		100.0	100.0			
Manila hemp	22.0	4.9									
Bl. Bol. Croc. Asbestos	0.50 0.65	2.9	15.9	15.9	15.9						
Esparto	7.0					100.0					
A	1.12								100.0		
AA	0.87									100.0	
AAA	0.62										100.0

TABLE II. FILTER MAT PHYSICAL PROPERTIES

Filter mat	Wt/area g/cm^2	Thickness λ cm_a	Bulk density ρ_{FM} [b] g/cm^3	Fiber density ρ_f g/cm^3	Porosity ε
Type 5	0.0128	0.048	0.267	1.565[c]	0.829
N11	0.0288	0.112	0.257	1.787[c]	0.856
N13	0.0263	0.089	0.296	1.787[c]	0.835
N15	0.0265	0.122	0.217	1.787[c]	0.879
Esparto	0.0204	0.068	0.300	1.33	0.774
Visc. 1.5D	0.0274	0.095	0.288	1.51	0.809
Visc. 3.0D	0.0112	0.050	0.224	1.51	0.852
A	0.0180	0.071	0.254	1.25	0.797
AA	0.0185	0.075	0.247	1.25	0.802
AAA	0.0069	0.028	0.246	1.25	0.803

[a]. TAPPI T-411 Method.   [b]. From Hall (1965).   [c]. Weighted mean.

TABLE III. FILTER MAT PARAMETERS

Filter mat	Mat fiber diam. d_f cm $\times 10^4$	Porosity ε	Fiber fraction σ	Ratio viod to fiber fractions ε/σ	Distance between fibers 2h cm $\times 10^4$
Type 5	0.65	0.829	0.171	4.848	3.15
N11	0.65	0.856	0.144	5.944	3.86
N13	0.50	0.835	0.165	5.061	2.53
N15	0.65	0.879	0.121	7.264	4.72
Esparto	7.0	0.774	0.226	3.425	23.98
Visc. 1.5D	12.0	0.809	0.191	4.236	50.83
Visc. 3.0D	17.0	0.852	0.148	5.757	97.87
A	1.12	0.797	0.203	3.926	4.40
AA	0.87	0.802	0.198	4.051	3.52
AAA	0.62	0.803	0.197	4.076	2.53

TABLE IV. DOP AEROSOL PROPERTIES

DOP Aerosol Diameter μ	Diffusivity D $cm^2 sec^{-1}$ $\times 10^6$	Relaxation Time τ sec $\times 10^7$
0.26	1.51	2.04
0.28	1.36	2.36
0.30	1.23	2.71
0.32	1.13	3.08

greatest discrepancies in both orders of magnitude and trends occur for k_I and k_D.

The reasons for these discrepancies are believed to be the following. Recall that one of the purposes of our study was the development of a simple quantitative basis for comparison and assessment of fibrous filters. General reviews [8,9] of aerosol filtration theory show that the log percent penetration should be the form

$$\log P = (b_1 V^{-2/3} + b_2 V^{-1}) + (b_3 V^{-1/2}) + (b_4 V) + (b_5) \quad (22)$$

if all ramifications of inertia, diffusion and interception are included. The terms in the first set of parentheses are for the diffusion mechanism. The term in the next set is for combined interception-diffusion. The term in the next set is for inertial impaction, and the one in the last set is for interception. Some of the coefficients, b_i, are available from tabulated results of calculations. Others exist as closed form expressions.

Davies[8] gives some useful tables from which we find that the b_2 term is negligible compared to the b_1 term for our systems and conditions. However, the other terms are not. Thus, the approximate equation, as judged by linear interpolation and extrapolation, applicable to our systems is the relation

$$\log P \simeq b_1 V^{-2/3} + b_3 V^{-1/2} + b_4 V + b_5. \quad (23)$$

As mentioned earlier some researchers have found "best fits" to their data using only the b_1 term (i.e. $V^{-2/3}$) and others using only the b_3 term (i.e. $V^{-1/2}$) in the Dorman procedure. Notice that equation (23) can be factored in two different ways i.e.

$$\log P \simeq b_1 V^{-2/3} \left(1 + \frac{b_3}{b_1} V^{1/6}\right) + b_4 V + b_5 \quad (24)$$

and

$$\log P \simeq b_3 V^{-1/2} \left(1 + \frac{b_1}{b_3} V^{-1/6}\right) + b_4 V + b_5 \quad (25)$$

The terms in parentheses are each mild functions of V for our range of values. However, in our calculation of Dorman parameters for Table VI we, like others, have neglected even this mild V dependence. Neglect of this is most serious when (d_p/d_f) is large. To use the full form (23) in the Dorman procedure is impossible if V_O is not determinable visually from penetration-velocity profiles. The foregoing considerations account for the non-monotonic behavior of k_D in Table VI.

The discrepancies in the experimental and calculated k_I values are attributable to the simple linear dependence on V of the b_4 inertial term. Conceptually, as V approaches infinity (large magnitude) the inertial removal must not approach infinity. Plots of the inertial contribution to capture made by point to

TABLE V. DOP PENETRATION AS A FUNCTION OF VELOCITY AND AEROSOL SIZE

Lin. Vol. cm/sec	Aerosol Diam. μ	Type 5	DOP% Penetration of									
			N11	N13	N15	Esparto	Visc. 1.5D	Visc. 3.0D	A	AA	AAA
7.2	0.26	29	0.006	0.092	0.004	31	72	87	5.5	0.009	0.022
	0.28	32	0.004	0.033	0	29	74	85	3.2	0.006	0.015
	0.30	27	0.001	0.030	0	32	75	88	2.7	0.002	0.004
	0.32	19	0	0.015	0	24	74	86	1.5	0	0
10.7	0.26	31	0.009	0.089	0.006	33	74	87	6.0	0.019	0.038
	0.28	33	0.005	0.030	0.001	30	76	86	4.2	0.012	0.021
	0.30	28	0.003	0.027	0	33	80	90	3.6	0.004	0.006
	0.32	21	0	0.010	0	24	75	88	1.7	0	0
14.2	0.26	31	0.011	0.085	0.007	33	74	87	7.0	0.026	0.048
	0.28	33	0.005	0.025	0.002	30	78	86	4.8	0.016	0.027
	0.30	29	0.003	0.022	0	33	82	90	4.1	0.005	0.007
	0.32	21	0	0.006	0	25	77	88	1.8	0	0
17.6	0.26	33	0.012	0.084	0.007	34	78	87	7.5	0.035	0.054
	0.28	33	0.005	0.024	0.003	30	80	87	5.3	0.018	0.040
	0.30	29	0.002	0.020	0	33	83	90	4.3	0.005	0.007
	0.32	21	0	0.005	0	26	80	88	1.8	0	0
26.7	0.26	33	0.014	0.074	0.008	33	78	87	8.0	0.052	0.072
	0.28	33	0.004	0.016	0.004	30	80	87	5.8	0.025	0.044
	0.30	29	0.001	0.012	0	32	83	90	4.8	0.005	0.006
	0.32	21	0	0.001	0	23	81	88	1.7	0	0
35.3	0.26	33	0.010	0.062	0.009	33	78	87	8.5	0.062	0.078
	0.28	33	0.002	0.009	0.003	28	80	87	6.0	0.028	0.044
	0.30	29	0	0.007	0	30	83	90	4.8	0.004	0.005

TABLE V. DOP PENETRATION AS A FUNCTION OF VELOCITY AND AEROSOL SIZE (CONTINUED)

Lin. Vol. cm/sec	Aerosol Diam μ	Type 5	N11	N13	N15	Esparto	DOP% Penetration of				
							Visc. 1.5D	Visc. 3.0D	A	AA	AAA
35.3	0.32	21	0	0	0	20	81	88	1.6	0	0
53.0	0.26	32	0.008	0.050	0.009	32	78	87	8.5	0.066	0.076
	0.28	32	0.001	0.005	0.002	26	80	87	6.0	0.027	0.038
	0.30	27	0	0.001	0	27	83	90	4.6	0.002	0.003
	0.32	19	0	0	0	14	81	88	1.4	0	0
71.0	0.26	30	0.006	0.029	0.007	28	78	87	8.5	0.058	0.074
	0.28	30	0	0.001	0.001	22	80	87	5.8	0.021	0.032
	0.30	25	0	0	0	23	83	90	4.0	0	0.002
	0.32	15	0	0	0	9	80	88	1.1	0	0
106.4	0.26	28	0.003	0.018	0.006	22	78	87	7.5	0.044	0.064
	0.28	27	0	0	0	16	80	87	5.0	0.012	0.025
	0.30	21	0	0	0	15	82	89	3.0	0	0
	0.32	11	0	0	0	4	78	80	0.6	0	0
141	0.26	24	0.001	0.008	0.003		78	83	6.5	0.028	0.040
	0.28	23	0	0	0		80	82	4.0	0.005	0.010
	0.30	17	0	0	0		80	86	1.6	0	0
	0.32	7	0	0	0			74	0.20	0	0

*DOP penetration of 0.000% shown as 0

TABLE VI. DORMAN PARAMETERS FOR FIBROUS FILTER MATS

Dorman Parameter from Penetration Data

Filter Mat	Inertial k_I (cm^{-2}sec)				Diffusional k_D (cm$^{-1/3}$sec$^{-2/3}$)				Interception k_R (cm^{-1})			
	0.26 μ	0.28 μ	0.30 μ	0.32 μ	0.26 μ	0.28 μ	0.30 μ	0.32 μ	0.26 μ	0.28 μ	0.30 μ	0.32 μ
Type 5	0.043	0.044	0.058	0.122	24.5	18.0	17.0	38.4	5.80	6.39	7.44	5.64
N11	0.123	0.204	0.741		42.2	9.68	85.9		26.3	33.8	15.5	
N13	0.110	0.423	0.568	0.934	11.1	50.0	52.2	37.8	30.7	24.5	22.9	26.9
N15	0.071	0.197			41.7	77.4			26.0	22.5		
Esparto	0.050	0.063	0.082	0.163	13.4	11.0	18.5	23.9	3.88	4.70	2.64	2.22
Visc. 1.5D			0.004	0.005			4.14	3.97			0.25	0.39
Visc. 3.0D		0.016	0.009	0.021		4.63	7.55	0.67		-0.04*	-0.38*	0.27
A	0.031	0.045	0.119	0.191	22.7	32.7	56.1	75.4	11.8	12.3	8.23	8.96
AA	0.092	0.158	0.329		82.7	89.7	76.4		31.7	32.6	39.8	
AAA	0.221	0.406	0.447		176	249	75.5	85.1	85.1	77.8	131	

*Negative value probably occurred due to an artifact present in the three data points available for calculation. Values are equivalent to zero since negative values have no physical significance.

TABLE VII.   CALCULATED DORMAN PARAMETERS FOR FIBROUS FILTER MATS

Dorman Parameter from Fuchs' Theory

Filter Mat	Inertial k_I (cm^{-2} sec)				Diffusional k_D (cm$^{-1/3}$ sec$^{-2/3}$)				Interception k_R (cm^{-1})			
	0.26 μ	0.28 μ	0.30 μ	0.32 μ	0.26 μ	0.28 μ	0.30 μ	0.32 μ	0.26 μ	0.28 μ	0.30 μ	0.32 μ
Type 5	1.03	1.19	1.37	1.56	53.0	49.4	46.2	43.6	13.5	15.6	18.0	20.4
N11	0.687	0.794	0.912		34.7	32.4	30.3		7.57	8.77	10.1	
N13	1.60	1.85	2.12	2.41	75.0	70.0	65.4	61.8	26.2	30.4	34.9	39.7
N15	0.459	0.531			22.7	21.4			4.25	4.93		
Esparto	0.018	0.021	0.024	0.027	2.02	1.89	1.76	1.67	0.03	0.03	0.04	0.04
Visc. 1.5D			0.005	0.006			0.47	0.44			0	0
Visc. 3.0D	0.001	0.001	0.002		0.15	0.14	0.13		0	0	0	
A	0.529	0.611	0.702	0.798	32.7	30.5	28.5	26.9	4.76	5.52	6.33	7.21
AA	0.826	0.955	1.10		46.9	43.8	40.9		9.34	10.8	12.4	
AAA	1.60	1.85	2.12		81.1	75.7	70.8		25.2	29.3	33.6	

point trajectory calculations using assumed flow fields show a
sigmoidal shape when plotted versus $\underline{V}$. One study [9,10] reproduced
this sigmoid shape with a relation in the form

$$\frac{a_1 V^3}{a_1 V^3 + a_2 V^2 + a_3} \tag{26}$$

Use of either $\underline{V}$ or $\underline{V}^2$ as the velocity dependence of the inertial
term is an approximation for this. Depending on the relative
magnitudes of the a_i it appears that $\underline{V}$ is more suitable as the
leading term in an expansion of (26). However, such an expansion
is only weakly justifiable. The use of the simple linear
dependence on $\underline{V}$ appears to be the cause of the discrepancies
between the experimental and calculated k_I.

Conclusion

Although not totally justifiable on theoretical grounds,
our modified Dorman procedure is a useful quantitative means to
compare filters. The additional computations of full flow theory,
although more satisfying conceptually, seem unwarranted for
practical applications. The full theory is, of course, necessary
for scientific studies.

Glossary

a_i numerical coefficients in approximate inertial term
b_i theoretical coefficients in log % penetration expression
d_f diameter of the aerosol filtering fiber (cm)
d_p diameter of the aerosol particulate (cm)
h one half the average distance between nearest neighbor
        filtering fibers (cm)
k_D Dorman parameter for diffusional filtration ($cm^{-1/3} sec^{-2/3}$)
k_R Dorman parameter for inertial filtration (cm^{-1})
k_I Dorman parameter for interceptional filtration ($cm^{-2} sec$)
ℓ mean free path of air molecules (cm)
m mass of aerosol particle (g)
x exponent of velocity in inertial term
y exponent of velocity in diffusion term
A numerical factor equal to unity in diffusion coefficient
D Diffusion coefficient of aerosol particles ($cm^2 sec^{-1}$)
L thickness of filter mat (cm)
No Avogadro's number (molecules $mole^{-1}$)
P Percent penetration of aerosols through a filter
P_a atmospheric pressure, (cm Hg)
R Universal gas constant (ergs $g^{-1} mole^{-1} deg^{-1}$)
T Absolute temperature (°K)

V   linear face velocity of flow through filter $(cm\ sec^{-1})$

V_m mean face velocity defined by egn(6) $(cm\ sec^{-1})$

α_D dimensionless Fuchs' coefficient for diffusional filtration

α_I dimensionless Fuchs' coefficient for inertial filtration

α_R dimensionless Fuchs' coefficient for interception

ε volume void fraction in filter mat

σ volume fiber fraction in filter mat

ρ_f density of fiber material $(g\ cm^{-3})$

ρ_{FM} Bulk density of filter mat $(g\ cm^{-3})$

ρ_P density of DOP aerosol particle $(g\ cm^{-3})$

τ mechanical relaxation time owing to viscous forces (sec)

References

1. Jonas, L. A., Lochboehler, C.M., Magee, W. S., Environ. Sci. Technol, **6**, 821 (1972).
2. Magee, W. S., Jonas, L. A., Anderson, W. L., Environ. Sci. Technol, **7**, 1131 (1973).
3. Dorman, R. G., "Aerodynamic Capture of Particles," Pergamon Press, Oxford, 1960.
4. Dorman, R. G., Air Water Pollution, **3**, 112 (1960).
5. Dorman, R. G., Chapter VIII in "Aerosol Science," C. N. Davies, Ed., Academic Press, New York, New York, 1966.
6. Fuchs, N. A., "The Mechanics of Aerosols," Pergamon Press, Macmillan, New York, New York, 1964.
7. Green, H. L. and Lane, W. R., "Particulate Clouds: Dusts, Smokes, and Mists," E. &F. N. SPON Ltd, London, 1957.
8. Davies, C. N., "Air Filtration," Academic Press, New York, New York 1973.
9. Pich, J., Chapter IX. in "Aerosol Science," C. N. Davies, Ed., Academic Press, New York, New York, 1966.
10. Landahl, H. and Hermann, K., J. Colloid Sci. **4**, 103 (1949).

9

Removal of Sulfur Dioxide in Stack Gases

DONALD A. ERDMAN

Potomac Electric Power Co., 1900 Pennsylvania Ave., Washington, D. C. 20006

In the late 60's when standards for sulfur content in fuel or emission limits were being promulgated the Potomac Electric Power Company (PEPCO) reviewed available fuel supplies and reached the conclusion that sufficient low sulfur fuel was not available to provide a reliable supply. PEPCO presented this position to the State of Maryland where our three largest base load stations are located. PEPCO stated that we would actively pursue stack gas desulfurization if we could continue to burn higher sulfur fuel while developing desulfurization. Maryland accepted this proposal.

We reviewed the various concepts for stack gas desulfurization and concluded that if at all possible the selected process should be a recovery system. This was not based so much on anticipating a profit from the sale of elemental sulfur or sulfuric acid, as on the lack of sufficient disposal areas near our plant sites to dispose of the wastes from a throw-away system. For instance, a lime throw-away system on 4 units at Dickerson would generate about 800,000 tons of sludge per year. A recovery system will produce about 180,000 tons of sulfuric acid per year. If a large percentage of power plants elect to produce sulfuric acid then disposal of acid can become a problem.

In agreement with the State, our first venture was a joint undertaking with Baltimore Gas & Electric in a Wellman-Lord process to be installed at B G & E's Crane Station. This process went into operation in April, 1969 and by early 1970 it was determined that the process as then constituted was not suitable for our use on a coal fired power boiler. This was not the same process as the one currently being promoted by Wellman-Lord which has demonstrated success on oil fired units.

In late 1970 PEPCO decided to proceed with the installation of a Chemico-Basic Mag-Ox scrubber on our Dickerson Unit No. 3. In 1971 the State held hearings on our delay in scrubber installation. Our participation with Baltimore Gas & Electric was reviewed. As a result of these hearings Maryland, in September,

1971, issued a compliance order that allowed us to burn fuel
with up to 2-1/4% sulfur provided we proceeded with the Chemico-
Basic installation with a target in-service date of June, 1973.
Actual start-up was September 13, 1973, just over three months
late. This delay was attributed in part to design changes based
on Chemico's experience at Boston Edison with a similar unit on
an oil fired boiler and on delays in equipment delivery.

The system consists of a 2-stage scrubber installed on our
Unit 3 and sized to take half the flue gas - 295,000 ACFM - equivalent to approximately 100 MW. The first stage of the scrubb-
is for particulate removal and the second stage for sulfur oxide
removal. The system is designed to take flue gas either ahead
of or after the existing electrostatic precipitators. The
scrubbing system is in parallel with the existing plant I.D.fans,
with a wet I.D. fan following the scrubber with this arrangement
loss of the scrubber system does not cause any loss of load.

The flue gas enters the first stage scrubber where it is
saturated and cooled from 250 F to 120 F. A bleed stream from
the first stage recycle flow carries flyash over to the thick-
eners for ultimate disposal to a settling pond. The cleaned and
saturated flue gas then enters the absorber through a fixed
throat venturi where it is contacted with the recyling MgO solution. The gas then goes to the wet I. D. fan, then through
mist eliminators to the stack where it mixes with the unscrubb-
ed gas from the other half of Unit 3. As magnesium sulfite
crystals build up in the slurry system a bleed stream is sent to
the centrifuge. The centrifuge cake is sent through the dryer
and the dry magnesium sulfite stored in a silo for ultimate
shipping to the reprocessing plant. The mother liquor from the
centrifuge is sent back to the absorber and this liquor is also
used to slake the MgO.

From September 13 to January 14 the scrubber was available
for about 27% of the time versus a boiler availability in excess of 95%. The problems during this initial operation were
various, corrosion leaks due to non-specified material in expansion joints, bearing failures in the sulfite handling
equipment, etc. The major problem though was in slaking the MgO.
There were continual problems with plugging in the MgO mix tank
and suction lines to the MgO make-up pumps.

In January the boiler was taken out of service for its
annual maintenance. During this outage a thorough inspection of
the system was made after some 700 hours' operation. Basically,
the system was in good shape. There was absolutely no scaling
or build-up in either scrubbing stage or on the I.D. fan or
demisters. However, in the first stage where the operating pH
is less than two, we had corrosion of nuts, bolts, hanger rods,
bellows, and the vessel itself. It was determined that the
corroded parts were of non-specified material. This emphasizes
the need to maintain quality control over the installation to
ascertain that when a specific grade of stainless steel is

specified that that grade is in fact installed. The corrosion
of the vessel occurred where the protective flake glass lining
was penetrated. The problem here was in some part due to improper application and, in part, due to construction damage after
the lining was installed. This was a very small percentage of
the total flake glass lining.

During this outage Chemico modified the MgO feed system
with the addition of a pre-mix tank. The system was finally
ready for service and started up on April 15. The intent was to
operate to verify the modifications and then shut down to set up
and operate for performance testing. The pre-mix tank improved
slaking but not to an acceptable standard for long-term operation.
We decided the end of April that the system could be operated
for performance testing. We shut down, checked inventory of MgO
and storage space in the magnesium sulfite silo, and found we
did not have enough MgO to run the test nor storage space for
sulfite. Boston Edison was using the reprocessing plant which is
located at an Essex sulfuric acid plant in Rumford, Rhode Island.
PEPCO would not have access to this facility until July 1, 1974.

In conjunction with Chemico, Basic, Essex and EPA, we are
embarked on a six-month test and demonstration period to run from
July to December, 1974. Due to lack of MgO we did not operate
in July. Chemico decided to try and improve the MgO slaking by
replacing the pre-mix tank with a "solids liquid mixing eductor."
This eductor was unsatisfactory and after one day's use was
replaced with the pre-mix tank. The pre-mix tank had been
modified by relocating baffles to improve agitation. Operation
with these modifications has been satisfactory. Virgin MgO was
received and operation resumed August 1st. Preliminary tests
indicated an SO_2 removal efficiency in the 70% range. Chemico
felt that the pressure drop across the absorber would have to be
increased to obtain the design 90% removal. These modifications
were made and the indicated removal efficiency on virgin MgO is
in excess of 90%. Our first recycle MgO was received and introduced into the system on August 16. This necessistated some
changes in operating procedures, such as temperature in the MgO
tank and dryer operating temperature. We are into the test program and although results are not currently available, it looks
encouraging.

Where does PEPCO stand after almost six years and approximately $9,000,000? Last January when we agreed to give this paper
we believed we would have more of the answers by now. Due to
delays, one of the main delays being access to the reprocessing
facility, we are currently in the middle of the performance
optimization and reliability testing. The system looks promising.
To date, there has been no problem to rule out the technical
feasibility of the MgO process. The area that still needs
verification is the cycling of MgO through the scrubber and
calciner for several cycles. This is necessary to assure that the

MgO can be reused and that impurities that may build up from
flyash carryover do not interfere with the production of a mark-
etable sulfuric acid. This data is also needed to determine the
amount of waste streams, if any, and also to determine system
economics.

PEPCO has a Certificate of Convenience and Necessity from
the State of Maryland to install an 800 MW addition at Dickerson. A condition of this certificate requires stack gas desulfurization on the existing three units and the new unit. The
allowable Federal ground level concentration for SO_2 is .030 PPM
on an annual basis or .14 on a 24-hour basis. Maryland has
taken the position that one source should not be allowed to contribute up to the maximum. They have therefore limited our
contribution to one-half the allowable ground level concentration
or approximately one-fourth the EPA health related standard.

I might add here that as part of our compliance order, we
are required by the State to maintain at each plant a seven days'
supply of fuel containing less than 1% S. This fuel is to be
used, when ordered by the State, in the event of area pollution
problems. The State has not found it necessary to order us to
switch to the low S fuel. This seems to indicate that SO_2 is
not a major contributor to pollution in our area.

We do not at this time have a definitive estimate of the
cost to the customer of meeting these requirements. However,
our rough estimates are that if we installed scrubbers at our
three Maryland plants, it could result in about a 20% increase
in the cost of electricity.

One of the big debates is are scrubbers reliable? The
three generating units at Dickerson have demonstrated an availability of 90% or better, since their installation. A scrubber
system should be capable of matching boiler availability. The
availability of the Dickerson scrubber continues to show improvement. It was 43.5% for the month of August and from August
13 to the end of the month the availability was 57%. August 13
is selected as that was the start-up date after the last modifi-
cation-increasing pressure drop across the absorber throat.

PEPCO continues to analyze the status of various stack gas
desulfurization systems, the availability of low sulfur fuel and
any other method for meeting our obiligation to produce reliable
power with a minimal adverse impact on the environment.

10

Kinetics of Trace Gas Adsorption from Contaminated Air

LEONARD A. JONAS, JOSEPH A. REHRMANN, and JACQUELINE M. ESKOW
Edgewood Arsenal, Aberdeen Proving Ground, Md. 21010

Abstract

A study was made of the kinetics of trace gas adsorption from contaminated air flowing into beds of activated carbon arranged in series. The purpose of the study was to determine if it were possible to predict the period of time for which the discharge flow, emitted to the atmosphere from two carbon filters in series, would not exceed an environmentally imposed limit on concentration which was below the sensitivity of existing monitoring equipment. The results of the study showed that (1) gas adsorption by carbon beds in series was equivalent to adsorption by a single bed with a proportionate increase in depth, (2) the present equations describing gas adsorption kinetics were applicable to carbon beds in series, (3) the gas concentration exiting the first carbon filter, monitored by a relatively insensitive alarm, could serve to predict the subsequent time period during which the concentration emitted from the second filter would never exceed the imposed emission standards.

Introduction

The study of trace gas removal from contaminated air was based upon experimentation and analysis of the adsorption characteristics of an activated carbon for the vapor isopropyl methylphosphonofluoridate (GB) under kinetic flow conditions. The carbon granules were packed to a reproducible bulk density in beds of

uniform cross-sectional area and subjected to constant inlet
vapor concentration, volume flowrate, and temperature but varying
bed weights. The breakthru time of the vapor through the bed, at
a known exit concentration, was determined for each bed weight, and
plotted as a straight line in accordance with existing adsorption
kinetics theory. Adsorption characteristics of the carbon were
determined from the slope and intercept of these straight line
curves.

Theory

The modified adsorption kinetics equation, originally derived from a continuity equation of mass between the gas entering an
adsorbent bed and the sum of the gas adsorbed by plus that penetrating through the bed, can be shown in the form of (1)

$$t_b = \frac{W_e}{C_o Q} \ [W - \frac{\rho_B Q}{k_v} \ \ln \ (C_o/C_x)] \tag{1}$$

where t_b is the gas breakthrough time in minutes at which the concentration C_x appears in the exit stream, C_o the inlet concentration in g/cm^3 , Q the volumetric flow rate in cm^3/min, ρ_B the bulk
density of the packed bed in g/cm^3, k_v the pseudo first order adsorption rate constant in min^{-1}, W the adsorption weight in g, and
W_e the kinetic saturation capacity in g/g at the arbitrarily
chosen ratio of C_x/C_o. In equation (1) ρ_B is determined as a
property of the granular size of the adsorbent when filled by
gravity settling in a container column, and the parameters C_o, C_x,
and Q are set by the conditions of test. Values of t_b plotted as
a function of W yield a straight line curve from whose slope and
x-axis intercept the properties W_e and k_v can be respectively calculated.

Although the adsorption of a gas molecule by an active site
within the micropore structure of a carbon was conceived by
Hiester and Vermeulen (2) to proceed by second order kinetics it
was shown by Jonas and Svirbely (3) that for a physically adsorbed

vapor pseudo first order kinetics should prevail, with respect to
gas molecules, over the range of gas concentration breakthru
$0 \leq C_x/C_o \leq 0.04$. The derivation and form of equation (1) are
consistent with the concept of pseudo first order kinetics in
this exit to inlet concentration range.

The test of the applicability of equation (1) to this study
resulted from the need to divide the carbon bed W into

$$W = W_1 + W_2 + \ldots + W_n \tag{2}$$

as sequential beds, and to expect that the invariance of the inlet
concentration C_o applied only to W_1. Thus, the physical make-up
of the carbon bed, with n=3, was shown as follows:

Experimental

Materials. The adsorbate used in these tests was 98% pure
isopropyl methylphosphonofluoridate (GB) obtained from Edgewood
Arsenal, Aberdeen Proving Ground, Maryland. It had a molecular
weight of 140.10 g/mole, and at 25°C a liquid density of 1.089
g/cm^3, a vapor pressure of 2.78 Torr, and a refractive index of
1.3810. These properties were either obtained from the Handbook
of Chemistry and Physics, 50 Ed. 1970 or determined using standard laboratory procedures. From these data the maximum vapor concentration capable of existing in air at 25°C was calculated from
the ideal gas law in the form of

$$\frac{g}{v} = \frac{PM}{RT} \tag{3}$$

where P is the pressure in atmospheres of the vapor, M is the
molecular weight, R is the gas constant of 82.05 cm^3 atmosphere
mole^{-1} deg^{-1}, T the degree Kelvin, and g/v the maximum concentration in g/cm^3. Since the gas tests were run at an inlet concentration of 5 mg/l (5×10^{-6} g/cm^3), and the maximum vapor concentration at 25°C was 20.9×10^{-6} g/cm^3, the relative pressure of
the GB vapor was 0.24.

The activated carbon adsorbent used in the tests was a 12-30 mesh BPL grade, lot no. 7502, from Pittsburgh Activated Carbon Co., Pittsburgh, PA, having as internal surface area of 1000 m^2/g. The BPL grade activated carbon has approximately 70-75% of its area associated with pores less that 20 $\overset{o}{A}$ in diameter. The impregnated carbon was also a 12-30 mesh BPL grade activated carbon from Pittsburgh Activated Carbon Co., which had been impregnated with an aqueous solution (12.0% NH_3) buffered by 10.9% carbonate ion and containing 8.5% cupric copper, 4.1% chromate ion, and 0.3% silver nitrate.

Equipment. The kinetic adsorption tests were carried out with a vapor adsorption test apparatus which was constructed of glass tubing fitted together with standard taper and ballsocket joints. The apparatus had three functional sections, one for vapor generation, another for vapor adsorption by the carbon, and the third for the detection of vapor penetration of the carbon bed. Jonas and Svirbely (3) have described the apparatus in greater detail and show schematically the interrelation of its operating components. The vapor penetration of each of the beds was detected by flame ionization in an F & M 5750 gas chromatograph. Dual columns were used, each consisting of a 12-ft. length of 3/16 in. diameter stainless steel tubing packed with 15% Dow-Corning silicone (DC LSX-3-0295) on a Chromasorb T support.

Procedures. The airflow rates of the adsorption test apparatus were calibrated by a gas flow meter and the activated carbons were oven dried at 120°C for a minimum of 16 hr. prior to storage in a desiccator. The adsorbate vapor was generated by the controlled passage of pre-dried nitrogen over the liquid surface, and the desired vapor concentration was then obtained by diluting this vapor with an auxiliary source of pre-dried nitrogen. The nitrogen was dried by passage through a series of deep columns of Drierite. A flow of 4000 cm^3/min of the vapor-mitrogen mixture was directed into the reservoir of the apparatus and flow aliquots sampled and analyzed until the desired concentration of

5×10^{-6} g/cm^3 was obtained. Adsorbent beds were formed by
gravity settling of the carbon granules in cylindrical glass
sample holders, and packed to uniform bulk densities. Various
bed weights of the adsorbents were exposed to the established
gas concentration drawn into the carbon bed at a volumetric flow
rate of 2432 cm^3/min (a superfical linear velocity of 575 cm/min
since the inside area of the carbon holder was 4.23 cm^2) at 25°C.
The exit air streams were monitored continually by passage into
the 25 cm^3 gas sampling valve of the gas chromatograph. The
breakthru time t_b was denoted as the time when the exit stream
from a particular bed weight showed the presence of a gas concentration of 5×10^{-8} g/cm^3, equal to an exit to inlet ratio of
0.01. Thus, the t_b for W_1 occurred when C_x/C_o reached 0.01, the
t_b for $W_1 + W_2$ occurred when C_y/C_o reached 0.01, and the t_b for
$W_1 + W_2 + W_3$ occurred when C_z/C_o reached 0.01.

Results and Discussion

Experimental values of gas breakthru time t_b as a function
of carbon bed weight W were obtained for the adsorbate vapor GB
on both activated carbon and ASC whetlerite carbon adsorbents.
Linear regression analyses were made on the data using an HP-45
calculator to obtain the regression equations and coefficients of
correlation ([4]). Figure 1 shows the regression equation plotted
as smooth lines and the experimental values, from which they were
obtained, as individual data points. A high degree of confidence
for the functional relation $t_b = t_b$ (W) was shown since the coefficient of correlation for the activated carbon data points was
0.987 and that for the ASC whetlerite carbon 0.995.

By multiplying through on the RHS of equation (1) one obtains

$$t_b = \frac{W_e W}{C_o Q} - \frac{W_e}{C_o} \frac{\rho_B}{k_v} \ln (C_o/C_x) \tag{4}$$

in which form the regression equations can be used to calculate
the critical bed weight, the kinetic adsorption capacity, and the

Table 1. GB Vapor Breakthru of Carbon under Airflow Conditions:

Q = 2432 cm³/min; A = 4.23 cm²; V_L = 575 cm/min; T = 25°C

Carbon	Inlet conc. of test C_o (g/cm³) x 10⁶	Carbon bed weight (g)				Breakthru time t_b at C_x/C_o = 0.01 (min)	
		W_1	W_2	W_3	Total	Observed	Normalized*
Activ. Carbon	4.593	2.0			2.0	30.5	28.0
Activ. Carbon	4.593	2.0	2.0		4.0	81.0	74.4
Activ. Carbon	4.593	2.0	2.0	2.0	6.0	126.0	115.7
Activ. Carbon	5.187	3.0			3.0	44.5	46.2
Activ. Carbon	5.187	3.0	3.0		6.0	125.0	129.7
Activ. Carbon	5.591	4.5			4.5	88.0	98.4
Activ. Carbon	5.591	4.5	4.5		9.0	211.0	235.9
ASC whetlerite	4.382	2.0			2.0	32.0	28.0
ASC whetlerite	4.382	2.0	2.0		4.0	92.0	80.6
ASC whetlerite	4.382	2.0	2.0	2.0	6.0	149.0	130.6
ASC whetlerite	6.010	3.0			3.0	51.0	61.3
ASC whetlerite	6.010	3.0	3.0		6.0	107.0	128.6
ASC whetlerite	5.921	6.0			6.0	98.0	116.1
ASC whetlerite	6.441	4.5			4.5	67.5	87.0
ASC whetlerite	6.441	4.5	4.5		9.0	154.0	198.4

* Normalized to C_o = 5.0 x 10⁻⁶ g/cm³

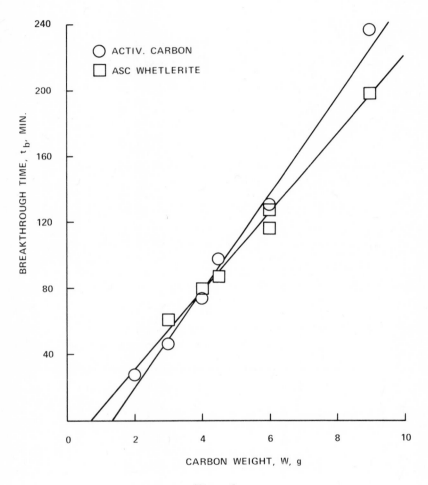

Figure 1.

adsorption rate constant under the experimental test conditions
prevailing. The conditions were as follows: temperature, 25°C;
internal cross-section of carbon holder, 4.23 cm^2; volume flowrate Q, 2432 cm^3/min; inlet concentration C_o, 5.0 x 10^{-6} g/cm^3;
exit concentration C_x (or C_y or C_z), 5.0 x 10^{-8} g/cm^3; and bulk
densities of packing, ρ_B, of 0.453 g/cm^3 for activated carbon and
0.482 g/cm^3 for ASC whetlerite carbon. By setting t_b of the
regression equation equal to zero and solving for W one obtains,
under the implied constraint that $W_e/C_o Q \neq 0$.

$$W = \frac{\rho_B Q}{k_v} \ln (C_o/C_x) \equiv W_c \tag{5}$$

where W_c is identified as the critical bed weight, or that weight
of carbon just sufficient to reduce C_o to C_x under the conditions
of test. The experimentally determined values of W_c, W_e, and k_v
for both carbons tested are shown in Table 2.

Comparison can be made between the adsorption parameters
obtained for GB vapor in this study, namely a W_e of 0.355 and a
k_v of 3712 for activated carbon and a W_e of 0.288 and a k_v of
7987 for ASC whetlerite, and values of 0.315, 20630, 0.324, and
21560 respectively reported in reference (1, a). In the latter
study, however, the inlet GB (shown by the symbols IMPF) vapor
was 5.85 x 10^{-7} g/cm^3, corresponding to a relative pressure at
25°C of 0.028, and the superficial linear velocity was 3001 cm/min.
In this study, however, the inlet concentration was 5 x 10^{-6} g/cm^3
(a relative pressure of 0.240) and the linear velocity 575 cm/min.
The major difference in results was manifested in the pseudo first
order adsorption rate constant. The increase in adsorption rate
constant with superficial linear velocity was shown by Jonas and
Rehrmann (5) to be sigmoidal in form, over the range 120 to 3001
cm/min, as expressed by the relation

$$k_v = \frac{a + b}{1 + \frac{a}{b} \exp [-(a+b)c \, V_L]} \tag{6}$$

Table 2. Adsorption Parameters for GB Vapor on Carbon

Carbon	Critical bed weight W_c (g)	Kinetic adsorption capacity W_e (g/g)	Adsorption rate constant k_v (min^{-1})	Kinetic equation
Activated carbon	1.367	0.355	3712	$t_b = 29.21\ W-39.93$
ASC whetlerite	0.676	0.288	7987	$t_b = 23.67\ W-16.00$

Although the values of the constants a,b, and c are determined by the system a determination of the extrema of equation (6) shows that $k_v \to b$ as $V_L \to 0$ and $k_v \to a + b$ as $V_L \to \infty$. The constant c determines the rate of change in k_v from the value b to a + b. The velocity of 3001 cm/min for a gas-air mixture through a 12-30 mesh carbon granular bed (mean granule diameter of 0.104 cm) showed a close approach to the maximum a + b value of k_v.

The general applicability of these relationships obtained with small gram quantities of carbon to multiple pound quantities, as well as the theory and method of calculating the parametric changes resulting from changes in temperature, are shown by Jonas et al (6) for the vapor dimethyl methylphosphonate, an established simulant in lieu of GB for the testing of activated carbon granules.

The high coefficient of correlation values obtained for the linear regression equations, shown in Table 2, wherein the values of breakthru time t_b were obtained in some cases for one bed of carbon, in some cases for two beds, and in others for three beds, and these values inserted at random and without singular designation into the regression analyses, indicated that the adsorption of GB vapor by multiple carbon beds, with intervening air plenums, was indistinguishable from that of one large bed equal to the sum of the individual beds.

An application of these relationships to the removal of gaseous contaminants in an exhaust control line occurs in those instances wherein the permissable emission limit is below the sensitivity of existing monitoring equipment. To overcome this difficulty the exhaust filter system can be constructed as two co-equal carbon filter beds separated by an air plenum. The trace gas monitoring device is positioned to detect gas penetrating the first carbon bed and entering the air plenum between the two carbon filters. By noting the time when penetration of the first filter occurred it is then possible to calculate that additional time during which the second filter will continue to filter the

exhaust without allowing the atmospheric discharge to exceed the
emission limit. Utilizing the information obtained in this study
on the activated carbon granules the following hypothetical situation can be established:

A leak concentration of 5×10^{-6} g GB per cm^3 of air
can be expected in an operation. The exhaust stack
for the operation contains two packed carbon bed filters
in series, separated by an air plenum. Each carbon
filter contains 1000 grams of 12-30 mesh activated
carbon granules. The volumetric flowrate through
the filters is 400 liters per min (4×10^5 cm^3/min).
The exhaust system operates at 25°C. The alarm or
monitoring device can detect a concentration of gas
as low as 100 nanograms per liter (10^{-10} g/cm^3), but
the emission limit is set at 1 nanogram per liter
(10^{-12} g/cm^3).

The stepwise procedure to be utilized for the calculations
necessary to assess the filter system is detailed as follows:

1. The time at which the monitoring device will alarm,
showing a penetration of GB vapor through the first filter of
100 n/l is 83.8 min. This value can be calculated from equation
(1), and is shown as

$$t_b = \frac{0.355}{(5 \times 10^{-6})(4 \times 10^5)} \left[\frac{1000 - (0.453)(4 \times 10^5)\ln\left(\frac{5 \times 10^{-6}}{10^{-10}}\right)}{3712} \right]$$

$$= 83.8 \text{ min}$$

2. The time required for the emission limit of 1 n GB/l to
penetrate both carbon filters and appear in the atmospheric air
will be 221.4 min.

$$t_b = \frac{0.355}{(5 \times 10^{-6})(4 \times 10^5)} \left[\frac{2000 - (0.453)(4 \times 10^5)\ln\left(\frac{5 \times 10^{-6}}{10^{-12}}\right)}{3712} \right]$$

$$= 221.4 \text{ min}$$

3. Thus, the filters can remain in operation for 137.6 min (221.4 - 83.8) after the monitoring device has alarmed to a penetration of 100 n GB/l through the first filter before the second filter will permit penetration of 1 n GB/l.

It is of interest to note that, in any such sequence of carbon bed filters, even though the gas concentration entering the first filter remains invariant the gas concentration penetrating this filter, and therefore entering each succeeding filter, will display a sigmoidal form with time. In fact, moreover, this concept can also be applied to any segmented portion of the first carbon bed filter. It is not, therefore, unreasonable to believe that only the first infinitesimal carbon layer sees an invariant inlet gas concentration which, having penetrated the antecedent layer, becomes the inlet concentration for the succeeding layer. As the flow velocity is increased the sigmoidal curve will approach the step function in form. This concept of gas flow movement through a bed of carbon is in full accord with the adsorption wave concept of Klotz (7).

References

1. a.   Jonas L. A. and Rehrmann J. A., Carbon 10, 657 (1972);
    b.   Jonas L. A. and Rehrmann J. A., Carbon 11, 59 (1973).
2.   Hiester N. K. and Vermeulen T., Chem. Eng. Progr.,
          48, 506 (1952).
3.   Jonas L. A. and Svirbely W. J., J. Catal., 24, 446 (1972)
4.   Hoel P. G., Introduction to Mathematical Statistics,
          p. 80, John Wiley and Sons, New York, N. Y. 1954.
5.   Jonas L. A. and Rehrmann J. A., Carbon 12, 95 (1974).
6.   Jonas L. A., Boardway J. C., and Meseke E. L., J. Coll.
          Interface Sci., manuscript accepted.
7.   Klotz I. M., Chem. Rev., 39, 244 (1946).

11

The Reaction Between Ozone and Hydrogen Sulfide: Kinetics and Effect of Added Gases

SOTIRIOS GLAVAS and SIDNEY TOBY

School of Chemistry, Rutgers University, New Brunswick, N. J. 08903

Abstract.

The reaction between O_3 and H_2S was studied from 20 to 70°C over a pressure range of 0.005 – 0.1 torr O_3 and 0.2 – 5 torr H_2S and the rate constant is given by log $(k/M^{-1/2} sec^{-1})$ = 5.0 ± 0.5 – (5000 ± 700)/2.30 RT. Up to 1.7 torr of added O_2 was found to have no effect on the reaction. However, 95 torr of added CO_2 reduced the observed rate constant slightly and 68 torr of C_2F_6 reduced the observed rate constant by a factor of 2. The results are explained in terms of a free radical mechanism which accounts for the observed product ratios found by us (3) and for the observed rate law. The effect of added gases on the reaction is also accounted for by the proposed mechanism. We use the data to extrapolate an approximate value for the initial bimolecular reaction between O_3 and H_2S. Implications for air pollution are considered.

Introduction.

Because of their importance as pollutants in the lower atmosphere, many reactants of ozone and of hydrogen sulfide are of great interest. The kinetics of their reaction with each other has been little studied and there has been no attempt to explain the observed rate law in terms of a mechanism. Cadle and Ledford (1) reported a rate law of $-d[O_3]/dt = k [O_3]^{3/2}$, i.e. zero order in H_2S. Hales, Wilkes and York (2) followed the rate by measuring the rate of production of SO_2 and found $d[SO_2]/dt = k [O_3]^{3/2} [H_2S]^{1/2}$. Both studies used flow systems with ozonized air or oxygen as a reactant and the stoichiometry was assumed to be $O_3 + H_2S \rightarrow SO_2 + H_2O$. A concerted molecular rearrangement is extremely unlikely and would not give rise to the observed rate laws. It is not clear whether the two reported rate laws are compatible with each other, and in any case no explanation has been given for the laws.

We have recently ($\underline{3}$) studied the reaction between O_3 and H_2S using pure O_3 as a reactant. We found the most abundant product was oxygen and that the (O_2 formed)/(O_3 used) ratio approached 1.5 as $[O_3]_0$ increased. In addition H_2O/SO_2 was found to vary considerably from unity. The rate of O_3 disappearance was found to obey a three-halves order rate law and two mechanisms were suggested which accounted for the observed kinetics and the observed product ratios. In this paper the mechanism will be considered in more detail and the effects of added inert gases will be discussed.

Product measurements and mass balances have been given elsewhere ($\underline{3}$) and will not be reported here.

Experimental Section.

A conventional high vacuum system was used with traps cooled with dry ice to exclude mercury vapor from the quartz reaction vessel. The cylindrical reaction vessel which was of length 21.9 cm and volume 460 cm^3, was connected to the reactant inlet and rest of the system by Teflon high vacuum stopcocks. An inlet tube consisting of a "cold finger" of approximately 5 cm^3 volume was attached via a stopcock to the reaction cell. The desired amount of H_2S was frozen into this inlet tube and allowed to warm before reacting.

The reaction vessel was mounted horizontally in a thermodatted oven ($\pm.8^o$) with quartz windows at each end. The ozone concentration was monitored by absorption of 254 nm radiation obtained from an Osram low-pressure mercury lamp and two interference filters. The decadic absorption coefficient of ozone was taken ($\underline{4}$) as 3010 $\underline{M}^{-1}$ cm^{-1} and checked by measuring the absorbance of O_3 which was then decomposed to O_2 and the pressure measured on a McLeod gauge. Beer plots were linear up to about 0.85 torr of O_3. O_3 pressures were kept below 0.1 torr because the reaction rate otherwise became too fast to measure with our system. The transmitted light beam was measured with a 1P28 photomultiplier connected to a Keithley 610A electrometer and a recorder.

Ozone was generated by passing oxygen (Matheson **Ultrapure** grade) through a discharge from a Tesla coil. The O_3 was condensed at -196° and the O_2 pumped away. Hydrogen sulfide, hexafluoroethane and carbon dioxide were each distilled before use, rejecting head and tail fractions. The initial pressures of reactants ranged from 0.005–0.1 torr for O_3 and 0.2–5 torr for H_2S.

Results.

The reaction was rapid with a typical half-life of ~2 secs. under the condition studied. Assuming a rate law of the form $-d[O_3]/dt = k[O_3]^m[H_2S]^n$ we studied the kinetics in the presence of excess H_2S for approximately 95% of the reaction. Log rate vs log $[O_3]_0$ plots gave slopes of between 1.5 and 2. We therefore used the integrated equations and made plots corresponding to orders of 1, 1.5, 2 and 2.5. The first and 2.5 order plots were distinctly curved but the 1.5 and second order plots were both straight with statistical correlation factors of more than 0.99 in most cases. However, the overall linearity of the 1.5 order plots was slightly better than that of the second order plots. The 1.5 order rate constants were measured over a 20-fold range of $[O_3]_0$ at 20, 50 and 70°C and the results were plotted in Figure 1. The rate constants show reasonably good independence of $[O_3]_0$. The effect of initial $[H_2S]_0$ on the 1.5 order rate constant at 20° is shown in Figure 2 and although there is scatter, there is no distinct trend. On the other hand, plots of rate constants of order 0.5 or higher in H_2S against $[H_2S]$ showed a clear dependence. We conclude that the rate law is zero order in H_2S and is given by

$$-d\,[O_3]/dt = k\,[O_3]^{1.5} \qquad (1)$$

Rate constants measured at 20, 50 and 70° were plotted as an Arrhenius plot which gave log $(k/M^{-1/2}\ sec^{-1}) = 5.0 \pm 0.5 - (5000 \pm 700)/2.3RT$.

The effect of added O_2, CO_2 and C_2F_6 was studied at 25° with initial ozone pressure kept at ca 0.03 torr and initial H_2S pressure kept at approximately 0.35 torr. These results are given in Table I.

TABLE I.   EFFECT OF ADDED GASES ON 1.5 ORDER RATE CONSTANT

Initial O_3, μ moles	Initial H_2S, μ moles	P_{CO_2}, Torr	$\dfrac{k \text{ at } 25°}{M^{-1/2}\ s^{-1}}$
0.94	7.66	0.12	40.4
0.65	8.64	0.14	59.0
0.60	8.68	0.65	59.9
0.51	8.70	1.14	45.2
0.74	8.62	2.00	45.5
0.68	7.96	2.04	62.5
0.60	8.60	10.6	63.3
0.75	8.58	28.1	55.9
0.66	8.60	41.5	52.0
0.64	8.53	49.8	49.9
0.82	8.67	69.0	43.2
0.79	8.52	87.0	43.0
0.65	8.10	95.4	41.6
		P_{O_2}	
0.65	9.00	0.35	46.6
0.68	7.86	0.90	41.4
0.51	8.15	1.30	41.4
0.74	8.69	1.68	51.3

TABLE I (cont'd.)

Initial O_3, μ moles	Initial H_2S, μ moles	$P_{C_2F_6}$ Torr	k, $\underline{M}^{-1/2}.sec^{-1}$
0.89	8.70	5.70	44.4
0.82	8.65	11.38	55.3
0.66	8.70	15.80	31.6
0.66	8.63	21.70	30.9
0.76	8.63	31.0	30.8
0.70	8.65	49.9	24.8
0.66	8.64	68.6	23.5

DISCUSSION

In order to explain the various product ratios found (3) and account for the observed rate law, we postulate two possible mechanisms differing in the initiating steps.

Mechanism A

$$H_2S + O_3 \xrightarrow{1} HSO + HO_2$$
$$HO_2 + O_3 \xrightarrow{2} HO + 2O_2$$
$$HO + H_2S \xrightarrow{3} H_2O + HS$$
$$HS + O_3 \xrightarrow{4} HSO + O_2$$
$$2HS + M \xrightarrow{5} H_2S_2 + M$$
$$HSO + O_3 \xrightarrow{6} HS + 2O_2$$
$$HSO + O_3 \xrightarrow{6a} HO + SO + O_2$$
$$SO + O_3 \xrightarrow{7} SO_2 + O_2$$

Taking steady states in $[HO]$, $[HO_2]$, $[HS]$, $[SO]$ and $[HSO]$ and putting $M = H_2S$ leads to

$$-d[O_3]/dt = (3+K)k_1[H_2S][O_3] + (2+K)k_1^{1/2} k_4 k_5^{-1/2}[O_3]^{3/2} \quad (2)$$
$$-d[H_2S]/dt = (2+K)k_1[H_2S][O_3] + K k_1^{1/2} k_4 k_5^{-1/2}[O_3]^{3/2} \quad (3)$$
$$d[H_2O]/dt = (1+K)k_1[H_2S][O_3] + K k_1^{1/2} k_4 k_5^{-1/2}[O_3]^{3/2} \quad (4)$$
$$d[SO_2]/dt = K k_1[H_2S][O_3] + K k_1^{1/2} k_4 k_5^{-1/2}[O_3]^{3/2} \quad (5)$$

where $K = k_{6a}/(k_6 + k_{6a})$. Eqs. (4) and (5) give

$$\frac{[H_2O]}{[SO_2]} = \frac{1 + K + k_1^{-1/2} k_4 k_5^{-1/2} K [O_3]^{1/2} [H_2S]^{-1}}{K + k_1^{-1/2} k_4 k_5^{-1/2} K [O_3]^{1/2} [H_2S]^{-1}}$$

Eqs. (3) and (5) give

$$\frac{[H_2S \text{ used}]}{[H_2O \text{ formed}]} = \frac{2 + K + k_1^{-1/2} k_4 k_5^{-1/2} K [O_3]^{1/2} [H_2S]^{-1}}{1 + K + k_1^{-1/2} k_4 k_5^{-1/2} K [O_3]^{1/2} [H_2S]^{-1}}$$

These product ratios have been found to be in qualitative agreement with experiment (3).

Steps 1 and 2 in the mechanism can be replaced by a sequence involving HSO_2 rather than HO_2. In additon, HSO_2 may also be formed by the reaction between HSO and O_3. We postulate, as an alternative possibility:

Mechanism B

$$H_2S + O_3 \xrightarrow{1a} HSO_2 + HO$$
$$HSO_2 + O_3 \xrightarrow{8} HSO + 2O_2$$
$$HSO + O_3 \xrightarrow{6b} HSO_2 + O_2$$

together with steps 3, 4, 5, 6, 6a and 7 of Mechanism A.

Mechanism B gives identical rate laws for eqs. (3), (4) and (5) with k_1 replaced by k_{1a}. In the case of eq. (2) the rate law is formally unchanged but K is now redefined as $K_1 = (k_{6a} + 2k_{6b})/(k_6 + k_{6a})$. Our data do not distinguish between Mechanisms A and B.

At the higher values of initial O_3 used a catalytic destruction of O_3 occurred. This requires a chain component of the mechanism. A likely possibility is step 2 along with $HO + O_3 \xrightarrow{9} HO_2 + O_2$. Evidence for the chain decomposition of O_3 resulting from steps 2 and 9 has recently been obtained by DeMore and Tschuikow-Roux (5). However, other chain sequences are possible, such as steps 6b and 8.

We did not detect H_2S_2 in our GC analyses (3) but Gunning et al (6) have reported that this substance readily decomposes: $H_2S_2 \rightarrow H_2S + S$. This would account for the sulfur we found deposited and the slightly low mass balance for S. According to Rommel and Schiff (7) at very low pressures the likely fate for HS radicals is $2HS \rightarrow H_2S + S$. However, in the higher pressures of our system step 5 is the most likely reaction.

As Table I shows, added O_2 up to a 60-fold excess relative to O_3, had no effect on the rate. This contrasts with what has been found for the reaction between O_3 and alkenes, where the radical intermediates are easily scavenged by O_2 leaving the residual initiating step (8).

In the presence of sufficient added deactivating gas M, eq. (2) becomes

$$-d[O_3]/dt = k_{10}[H_2S][O_3] + k_{11}[O_3]^{3/2}[H_2S]^{1/2}/([H_2S] + \beta[M])^{1/2} \quad (6)$$

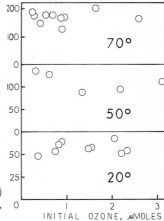

Figure 1. Plot of 3/2 order rate constant (M⁻¹ᐟ² sec⁻¹) vs. initial O_3. H_2S ~ 25, 8, and 7.5 μmoles at 20, 50, and 70°.

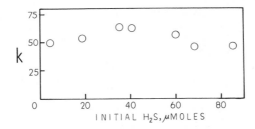

Figure 2. Plot of 3/2 order rate constant ($M^{-1/2}$ sec⁻¹) vs. initial H_2S at 20°. Initial $O_3 = 0.75$ μmoles.

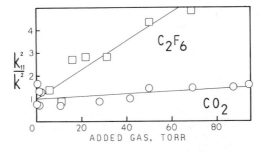

Figure 3. Effect of added deactivators. Plot of k_{11}^2/k^2 vs. pressure of added gas at 25° with initial $O_3 = 0.75$ μmoles, initial $H_2S = 8.5$ μmoles.

where $k_{10} = k_1(3+K)$, $k_{11} = k_1^{1/2}k_4k_5^{-1/2}(2 + K)$ and β is the relative deactivating efficiency of M. If we compare the observed 3/2 order rate constant k in eq. (1) with k_{11} in eq. (6) we may write

$$k_{11}^2/k^2 = 1 + \beta[M]/[H_2S] \qquad (7)$$

A plot of k_{11}^2/k^2 vs [M] for CO_2 and C_2F_6 from the data in Table I is shown in Figure 3. From the slope we obtain $\beta(CO_2)$ = 1.9 x 10^{-3} and $\beta(C_2F_6)$ = 2.2 x 10^{-2}. These values indicate the relative deactivating efficiencies of the added species in the dimerization of HS radicals.

If the second term on the right in eq. (2) predominates over the first then a three halves order rate law results for the rate of disappearance of ozone. Our observed rate constant is then equal to $k_1^{1/2}k_4k_5^{-1/2}(2 + K)$ for Mechanism A or $k_{1a}^{1/2}k_4k_5^{-1/2}$ $(2 + K_1)$ for Mechanism B. We may then compare our observed rate constant of $\log(k/M^{-1/2}sec^{-1})$ = 5.0 $\pm$ 0.5 - (5000 $\pm$ 700)/2.30 RT and our value of $k(25°)$ = 52 $\pm$ 8 $M^{-1/2}sec^{-1}$, with that of Cadle and Ledford (1) who reported $\log(k/M^{-1/2}sec^{-1})$ = 6.9 - 8300/2.30 RT and $k(25°)$ = 8.5 $\pm$ 2 $M^{-1/2}$ sec ^{1}. They used a flow system and found some heterogeneity, and in view of the large differences in experimental methods used, the agreement is encouraging. Hales, Wilkes and York (2) found $d[SO_2]/dt = k[H_2S]^{1/2}[O_3]^{3/2}$ using an air carrier gas. According to our mechanism eq. (5) would then become $d[SO_2]/dt = k[H_2S]^{1/2}[O_3]^{3/2}/[Air]^{1/2}$ and at constant total pressure their rate law is then compatible with the one reported here.

We could not measure the second order rate constant in eq. (2) directly but two methods were employed to obtain indirect estimates. Method I consisted of taking the limiting slopes of pseudo first order plots at long reaction times. As $[O_3]$ decreases the term in eq. (2) which is first order in O_3 will eventually predominate over the 1.5 order term. In Method II the rate was plotted against $[O_3]$ and the tangent at the origin will be governed mostly by the term which is first order in O_3. The rate constants evaluated by the two methods at 25, 50 and 70° are given in Table II.

TABLE II. EVALUATION OF SECOND ORDER RATE CONSTANTS
BY METHODS I AND II

$[O_3]_0$ $\underline{M}$ X 10^6	$[H_2S]_0$ 25°	k(I) $\underline{M}^{-1}sec^{-1}$ X 10^{-3}	k(II)
2.27	37.0	1.6	1.6
1.07	37.0	2.0	2.7
0.64	27.0	1.6	0.3
2.97	37.0	0.65	3.5
2.01	37.0	0.57	1.4

TABLE II (cont'd.)

$[O_3]$ M	X	$[H_2S]$ 10^6	$k(I)$ $M^{-1}sec^{-1}$ X	$k(II)$ 10^{-3}
		50°		
2.22		34.7	1.7	2.0
2.95		35.5	1.5	1.7
0.98		38.8	2.6	4.1
1.49		36.8	1.2	4.9
0.43		34.0	2.6	1.8
		70°		
2.06		16.8	4.9	7.5
1.92		16.8	4.8	8.3
1.46		16.8	6.8	7.4
1.20		15.5	5.9	7.4
0.63		16.4	3.42	6.0
0.57		16.5	5.5	5.2
0.91		15.9	4.9	5.9

Although the points show considerable scatter, the agreement between the two methods is fair and we make the approximate estimate that $\log (k/M^{-1}sec^{-1}) = 8.1 \pm 0.4 - (6800 \pm 600)/2.3RT$. This rate constant may be set equal to $k_1(3 + K)$ for Mechanism A or $k_{1a} (3+K_1)$ for Mechanism B. K is a fraction, and so is K_1 if we make the assumption that $k_{6b} < (k_6 + k_{6a})$ based on crude bond energy considerations. Thus, for either mechanism the initiating rate constant is ~ 500 $M^{-1}sec^{-1}$ at room temperature. This may be compared with the upper limit of 12 $M^{-1}sec^{-1}$ given by Becker, Inocencio and Schurath (9) and we cannot account for this discrepancy.

The initiating step in Mechanism A or B presumably takes place via a 5-membered ozonide ring:

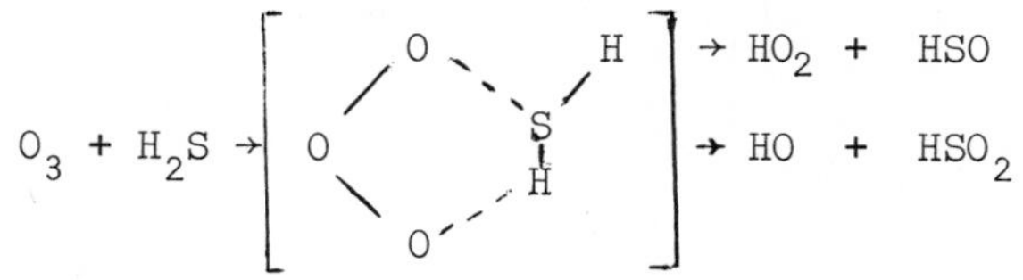

This transition state is similar to the cyclic transition state proposed by DeMore for the reaction between O_3 and ethylene (10). DeMore calculated that a 5-membered ring would give an A-factor of ~ 10^8 $M^{-1}sec^{-1}$. This value is similar to our A-factor for the initiating step and correlates with the fact that $S°(H_2S) = 49.2$ e.u. and $S°(C_2H_4) = 52.4$ e.u. are comparable. The overall reaction between O_3 and H_2S is chemiluminescent and Pitts et al (11) have attributed the emission to SO_2^*. This is consistent with the high exothermicity (106 k cal) of step 7. A more detailed study by Becker et al (9) of the chemiluminescence found additional bands which were tentatively identified as HSO^* or HSO_2^*.

Implications for Air Pollution.

Hydrogen sulfide is not normally an important atmospheric pollutant. By way of comparison, typical SO_2 concentrations in urban atmospheres are ~ 0.2 ppm by volume (12) but H_2S values are much lower, perhaps 0.01 ppm (although the analytical procedures used have been criticized (13)). From our work the rate law for the reaction between H_2S and O_3 in air may be written $-d[O_3]/dt = k_1[H_2S][O_3] + k[H_2S]^{1/2}[O_3]^{3/2}[Air]^{-1/2}$. Inserting our values for the rate constants at 25° and substituting typical values of $[Air] = 0.04$ $\underline{M}$, $[H_2S] = 5 \times 10^{-10}$ $\underline{M}$, and $[O_3] = 2 \times 10^{-8}$ $\underline{M}$ gives the rate as 2×10^{-14} $\underline{M}$ sec^{-1}. This is five orders of magnitude smaller than the rate of a "fast" reaction such as $NO + O_3 \rightarrow NO_2 + O_2$ using typical concentrations with $[NO] \simeq [O_3]$.

As a result of their upper limit for the bimolecular rate constant, Becker et al (9) concluded that the oxidation of H_2S by O_3 is unimportant under atmospheric conditions. Although our results differ from theirs, our conclusion is the same. However, it must be emphasized that oxidation of H_2S is relatively rapid in aqueous solution. Thus, as Kellogg et al in a review of the sulfur cycle have pointed out (13), the oxidation of H_2S in the sea is probably an important sequence. In the atmosphere, however, the reaction would probably have to take place via atomic oxygen or possibly in aerosols in order to be of importance.

Acknowledgment. We thank the Biomedical Council, Rutgers University for support in this work.

Literature Cited.

1. R. D. Cadle and M. Ledford, Int. J. Air and Water Pollution, 10, 25 (1966).

2. J. M. Hales, J. O. Wilkes and J. L. York, Atmos. Environ., 3, 657 (1969).

3. S. Glavas and S. Toby, J. Phys. Chem., 78, 0000 (1975).

4. M. Griggs, J. Chem. Phys., 49, 858 (1968).

5. W. B. DeMore and E. Tschuikow-Roux, J. Phys. Chem., 78, 1447 (1974).

6. P. Fowles, M. DeSorgo, A. J. Yarwood, O. P. Strausz and H. E. Gunning, J. Am. Chem. Soc., 89, 1352 (1967).

7. H. Rommel and H. I. Schiff, Int. J. Chem. Kins., 4, 547 (1972).

8. F. S. Toby and S. Toby, Int. J. Chem. Kins., 7, 0000 (1975).

9. K. H. Becker, M. Inocencio and U. Schurath, Abstracts of CODATA Symposium, Warrenton, Va., Sept. 1974. To be published in Int. J. Chem. Kinetics

10. W. B. DeMore, Int. J. Chem. Kins., 1, 209 (1969).

11. W. A. Kummer, J. N. Pitts, Jr. and R. P. Steer, Envir. Sci. Technol., 5, 1045 (1971).

12. R. S. Berry and P. A. Lehman, Ann. Rev. Phys. Chem., 22, 47 (1971).

13. W. W. Kellogg, R. D. Cadle, E. R. Allen, A. L. Lazrus and E. A. Martell, Science, 175, 587 (1972).

12

Photolysis of Alkyl Nitrites and Benzyl Nitrite at Low Concentrations—An Infrared Study

BRUCE W. GAY, JR., RICHARD C. NOONAN, PHILIP L. HANST, and
JOSEPH J. BUFALINI

Chemistry and Physics Laboratory, Environmental Protection Agency,
Research Triangle Park, N.C. 27711

Abstract

Methyl, ethyl, propyl, and benzyl nitrite were photolyzed
at low concentration in air or oxygen with analysis of products
by long path infrared techniques. In all but the methyl nitrite photolysis the major product was a peroxyacyl nitrate type
compound. The photolysis of methyl nitrite gave nitric acid as
the major nitrogen containing product, with formaldehyde, carbon
monoxide, and formic acid as carbon containing products. The
carbon balance in all systems was good. The nitrogen balance
became poorer as percentage of PAN-type product decreased and
nitric acid increased. The poor nitrogen balance is ascribed
to nitric acid absorption on the reactor surface.

Introduction

Many of the interesting compounds that make up the urban
smog are formed by photochemical reactions of compounds emitted
into the atmosphere. The photooxidation of certain hydrocarbons
in the presence of oxides of nitrogen forms a class of compounds
that are strong oxidizing agents. The most abundant in this
class of compounds found in the atmosphere is peroxyacetyl nitrate (PAN). Recent studies suggest that a well-reacted atmosphere, i.e., stagnant atmosphere in late afternoon, contains
PAN as the major nitrogen containing product (1). Peroxy
propionyl nitrate (PPN) a homolog of the PAN series has been observed in the atmosphere along with PAN but at lower concentrations (2), (3). The presence of PAN type compounds is important.
They are responsible for severe plant damage (2), (4) and concentrations in the parts per billion level range can cause eye
irritation (5). Peroxybenzoyl nitrate (PBzN) has been shown to
be a much stronger eye irritant than PAN by two orders of
magnitude (5). Severe eye irritation during low or moderate smog
conditions may be a result of low concentrations of PBzN. Although PBzN has not been detected in the urban atmosphere, it has
been observed as a product in laboratory studies of surrogate

atmospheric mixes. It is possible that an increase in aromatic content in auto fuel could result in PBzN formation and detection.

The simplest homolog of PAN type compounds that contains only one carbon per molecule is peroxyformyl nitrate (PFN). Until recently PFN had never been observed although many had attempted to synthesize this compound. It has been postulated that PFN does form but quickly dissociates to nitric acid and carbon dioxide via a six membered ring configuration (6). Peroxyformyl nitrate is presently under study in this laboratory. Preliminary results indicate that it is created when formyl radicals are produced in one atmosphere of oxygen containing 30 parts per million of NO_2. The compound appears to decompose at room temperature with a half life of a few minutes.

PAN type compounds for laboratory studies have been prepared in the past by various methods. Stephens (7) photolyzed olefins in the presence of oxides of nitrogen and obtained PAN. The photolysis of diacetyl and nitrogen dioxide in air produces PAN (8). Tuesday (9) synthesized PAN in the dark reaction of acetaldehyde with NO_2 and ozone. Stephens (10) photolyzed alkyl nitrites in oxygen to obtain the compound. More recently Gay (11) photolyzed chlorine in the presence of acetaldehyde, NO_2, and oxygen to produce PAN.

This study was undertaken to investigate the products of photolysis of alkyl and benzyl nitrite at low concentration in a photochemical reaction cell with fourier transform infrared equipment. From product formation and reactant disappearance, the mechanism and kinetics of the reaction were determined.

Experimental

The irradiations of the compounds were carried out in a long cylindrical glass chamber of 6.18 meters length and 0.31 meter diameter. The chamber was constructed by connecting four borosilicate Corning QVF (PS 12-60) pipes. Aluminum spacers 1.91 cm wide with 0.32 cm teflon gaskets on either side were sandwiched between the pipe sections. Three such spacers having injection ports to introduce samples into the cell were used. The chamber ends were capped with 3.18 cm flat plexiglas endplates and 0.32 cm teflon gaskets. The internal volume of the chamber was 460 liters. A large displacement vacuum pump was connected to one end plate with a ball vacuum valve through which the chamber could be evaluated to a pressure less than one torr. Leak rate of the chamber was less than 50 torr per day. Samples were mixed and introduced into the cell through a glass manifold from a glass gas sample handling system including mixing bulbs of known volume, pressure gauges of 0-50 torr and 0 - 800 torr. Measured amounts of gaseous material were transferred from the known volume bulbs of the gas handling system into the chamber via the manifold and ports in the aluminum spacers. Surrounding each of the four chamber sections were cylindrical aluminum light banks in two sections. Each section contained eight GE-F40 BLB black light lamps with energy maxima of 3660 $\overset{o}{A}$. Running the length of each chamber section were sixteen lamps arranged and controlled to give uniform light intensity when all, one-half, or one quarter of the lamps were turned on (with one quarter of the lamps on $k_{d(NO_2)}$= 0.17 min^{-1}). Lamp ballasts (GE-6G 1020 B) were mounted on a heat sink away from the chamber. The light intensity, $k_{d(NO_2)}$ was measured by the photolysis of NO_2 in oxygen free nitrogen (9).

The chamber used for the photochemical reactions contained within it an eight mirror optical system (12). Thus, the chamber served also as a long path multiple reflection in-frared absorption cell capable of lengths of 43n meters, where n is an integer between 1 and 15. In these experiments path lengths of 172 and 344 meters were used. Potassium bromide entrance and exit windows were used on the cell. The infrared instrument was a Digilab Fourier Transform Spectrometer that uses a scanning Michelson interferometer with a KBr beam splitter coated with germanium. Liquid nitrogen cooled detectors were used to cover the spectral region from 700 to 3400 cm^{-1}. A more detailed discussion of the optical system, interferometer and spectral data techniques are given elsewhere (13). The ab-sorption coefficients of reactants and products were measured in most cases under conditions similar to those used in the ex-periment, i.e. 750-760 torr total pressure of tank oxygen or air, $30^{o}C$, low relative humidity (<1 percent), and one wave number resolution at 172 or 344 meters path length.

Initial nitrite concentrations ranged from 2 to 10 ppm. Concentrations from spectral data were calculated using the Beer-Lambert Law, $\ell n\ Io/I = kLp$, where Io and I are the in-cident and transmitted intensity, respectively, k is the ab-sorption coefficient, L the optical path length, and p the partial pressure of absorbing species.

All chemicals used to make the starting nitrite compounds were research grade and used without further laboratory purifi-cation. The nitrite compounds were prepurified just before irradiation. The zero grade dry tank air was low in total hydro-carbons and carbon monoxide (typically 0.5 - 1.5 ppm CH_4, 0.1 - 0.5 ppm CO); zero grade dry tank oxygen did contain 1.7 ppm methane.

Methyl nitrite was prepared from methyl alcohol and sodium nitrite under acid conditions (14). The gaseous methyl nitrite was collected at solid carbon dioxide temperature and redistilled in vacuum by room temperature heating and collection at liquid nitrogen temperature.

Propyl nitrite was prepared by a similar method using propyl alcohol in place of the methyl alcohol. The product was sepa-rated and purified using conventional vacuum distillation techniques.

Benzyl nitrite was prepared (15) by adding dropwise a 40 percent aqueous solution of $Al_2(SO_4)_3$ to a vigorously stirred solution of benzaldehyde, $NaNO_2$ and water over a period of two hours. After standing for an hour the supernatant layer was decanted and dried with anhydrous sodium sulfate. The benzyl nitrite was vacuum distilled to further purify the compound.

Ethyl nitrite was purchased from Mallinckrodt Chem. Corp. and contained eight percent ethanol. The ethyl nitrite mixture was vacuum distilled to obtain the pure compound.

Results

Methyl Nitrite. Methyl nitrite was irradiated in both zero tank air and tank oxygen without any noticeable change in product concentration or reactant half life. Figure 1 is a graph of concentration versus time for 5.5 ppm methyl nitrite irradiated in oxygen at 760 torr with light intensity of $k_{d(NO_2)} = 0.3$ min^{-1}. The half-life time for methyl nitrite at this light intensity is 9 min. The photolysis takes place as follows:

$$CH_3ONO + h\nu \rightarrow CH_3O + NO \tag{1}$$

The back reaction of (1) takes place and must be considered:

$$CH_3O + NO \rightarrow CH_3ONO \tag{2}$$

The methoxy radical undergoes a number of reactions which lead to the production of formaldehyde:

$$CH_3O + O_2 \rightarrow HCHO + HO_2 \tag{3}$$

$$CH_3O + NO \rightarrow HCHO + HNO \tag{4}$$

$$CH_3O + NO_2 \rightarrow HCHO + HNO_2 \tag{5}$$

The reaction to form methyl nitrate also becomes important at higher initial concentrations of methyl nitrite or when irradiated in an oxygen poor system.

$$CH_3O + NO_2 \rightarrow CH_3ONO_2 \tag{6}$$

At initial concentrations lower than about 8 ppm methyl nitrite in air no methyl nitrate was detected. The detection limit of methyl nitrate in the system was of the order of 0.04 ppm.

Carbon monoxide forms by a sequence of reactions involving formaldehyde.

$$CH_2O + OH \rightarrow HOH + CHO \tag{7}$$

$$CHO + O_2 \rightarrow HO_2 + CO \tag{8}$$

Formic acid forms in accordance with reactions listed below but at a slower rate than carbon monoxide.

$$CHO + O_2 \rightarrow H\overset{\overset{O}{\|}}{C}OO \tag{9}$$

$$2H\overset{\overset{O}{\|}}{C}OO \rightarrow 2H\overset{\overset{O}{\|}}{C}O + O_2 \tag{10}$$

$$H\overset{\overset{O}{\|}}{C}O + HCHO \rightarrow HCOOH + HCO \tag{11}$$

The formation of peroxyformyl nitrate could take place by reaction of NO_2 with the product in reaction (9).

$$H\overset{\overset{O}{\|}}{C}OO + NO_2 \rightarrow H\overset{\overset{O}{\|}}{C}OONO_2 \tag{12}$$

Nitric acid is observed to form as soon as the lamps are turned on. The principle reactions for its formation are the following:

$$NO_2 + OH \overset{M}{\rightarrow} HNO_3 \tag{13}$$

$$N_2O_5 + H_2O \overset{M}{\rightarrow} 2HNO_3 \tag{14}$$

Nitrogen pentoxide is an observed product in these systems that contain about 100 ppm water vapor. The reactions that give nitrogen pentoxide involve ozone and oxides of nitrogen.

$$NO + O_3 \rightarrow O_2 + NO_2 \tag{15}$$

$$NO_2 + O_3 \rightarrow O_2 + NO_3 \tag{16}$$

$$NO_2 + NO_3 \rightarrow N_2O_5 \tag{17}$$

Atomic oxygen formed by the photolysis of NO_2 adds to oxygen in a third body reaction to produce ozone.

$$NO_2 + h\nu \rightarrow NO + O \tag{18}$$

$$O + O_2 \overset{M}{\rightarrow} O_3 \tag{19}$$

There are of course many other competitive reactions that are taking place. The methyl nitrite system has been computer modeled using 55 reactions, 26 species, and the best available rate constants (adjustments to rates, where needed). The computer model used was similar to that of Hecht ([16]). A good fit to the experimental data was obtained with this model.

Ethyl Nitrite.   The photolysis of ethyl nitrite is shown in figure 2.   It photodissociates in a manner similar to methyl nitrite reaction (1), to give NO and an ethoxy radical.

$$CH_3CH_2ONO \overset{h\nu}{\rightarrow} CH_3CH_2O + NO \qquad (20)$$

The ethoxy radical can react with a number of different species to form acetaldehyde:

$$CH_3CH_2O + NO \rightarrow HNO + CH_3CHO \qquad (21)$$

$$CH_3CH_2O + NO_2 \rightarrow HNO_2 + CH_3CHO \qquad (22)$$

Acetaldehyde loses the aldehydic hydrogen by reaction with other species present to form the acyl radical:

$$CH_3CHO + OH \rightarrow HOH + CH_3CO \qquad (23)$$

The reaction with oxygen forms the peroxyacetyl radical, which can then react with nitrogen dioxide to form PAN:

$$CH_3CO + O_2 \rightarrow CH_3\overset{\overset{O}{\parallel}}{C}OO \qquad (24)$$

$$CH_3\overset{\overset{O}{\parallel}}{C}OO + NO_2 \rightarrow CH_3\overset{\overset{O}{\parallel}}{C}OONO_2 \qquad (25)$$

If it were not for the ozone and other species which react rapidly with NO to keep its concentration low, the peroxyacetyl radical would react with NO to oxidize it to NO_2.

$$CH_3\overset{\overset{O}{\parallel}}{C}OO + NO \rightarrow CH_3\overset{\overset{O}{\parallel}}{C}O + NO_2 \qquad (26)$$

The acylate radical that formed will decompose:

$$CH_3\overset{\overset{O}{\parallel}}{C}O \rightarrow CH_3 + CO_2 \qquad (27)$$

Propyl Nitrite.   The results of the photolysis of propyl nitrite in oxygen are shown in figure 3.   The photolysis was carried out at twice the light intensity of the previous experiments.   Thus, the decrease in concentration of propyl nitrite is more rapid.   Again, the initial reaction is similar to the photolysis of methyl nitrite, yielding NO and the propoxy radical.   By a series of reactions analogous to reactions 21 through 25, the major products, propionaldehyde and peroxypropionyl nitrate are formed.   Peroxypropionyl nitrate accounts for 62 percent of the nitrogen in the system.   The propion-

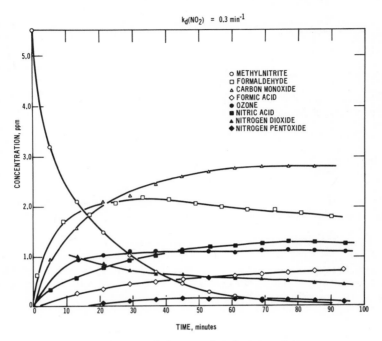

Figure 1.   Methyl nitrite photolysis in oxygen

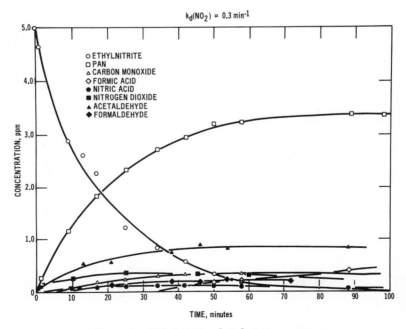

Figure 2.   Ethyl nitrite photolysis in oxygen

aldehyde and peroxypropionyl nitrite combined account for 95 percent of the carbon in the system.

Benzyl Nitrite. The photolysis of benzyl nitrite under conditions similar to those of methyl and ethyl nitrite photolysis is shown in Figure 4. Peroxybenzoyl nitrate and benzaldehyde account for 89 percent of the carbon in the reacted system. All of the nitrogen can be accounted for as peroxybenzoyl nitrate, NO_2, and a small amount of nitric acid. The concentration of NO_2 formed was substantially higher in this system as compared to the ethyl and propyl nitrite systems. No multicarbon products could be observed that were a result of ring rupture. However, small amounts of formic acid and carbon monoxide were observed. This suggests that at least some small fraction of the aromatic underwent ring fracture.

Infrared Spectral Data. Figures 5, 6, 7, and 8 show in the upper half, the infrared spectra of the organic nitrites before irradiation. The lower half of the spectra show the products formed after irradiation. These spectra were obtained for a 172 meter path length at 760 torr pressure. The resolution was degraded by a factor of four to produce these spectra of 4 cm^{-1} resolution.

Discussion

The photolyses of methyl, ethyl, propyl, and benzyl nitrite at low concentrations in oxygen or air were conducted to compare mechanisms of peroxyacyl nitrate formation. In all but the methyl nitrite photolysis the major product was a peroxyacyl nitrate. The long path infrared absorption technique would have been able to detect low concentrations of peroxyformyl nitrate had it formed and been stable. It has been postulated that PFN forms a six membered ring configuration which quickly splits to form nitric acid and CO_2 (6).

$$\rightarrow HNO_3 + CO_2$$

It it possible that PFN does form but because it quickly dissociates to HNO_3 and CO_2 its equilibrium concentration is much lower than our detection limit. Since a large amount of

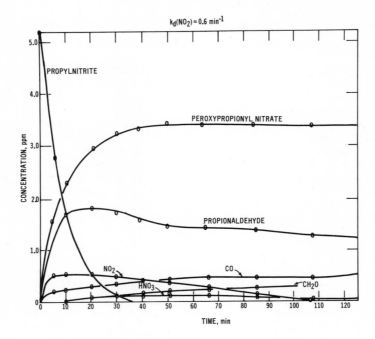

Figure 3.   *Propyl nitrite photolysis in oxygen*

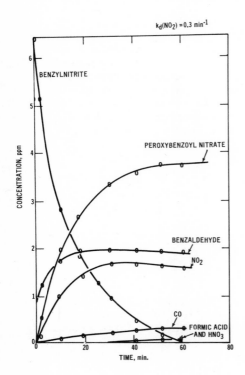

Figure 4.   *Benzyl nitrite photolysis in oxygen*

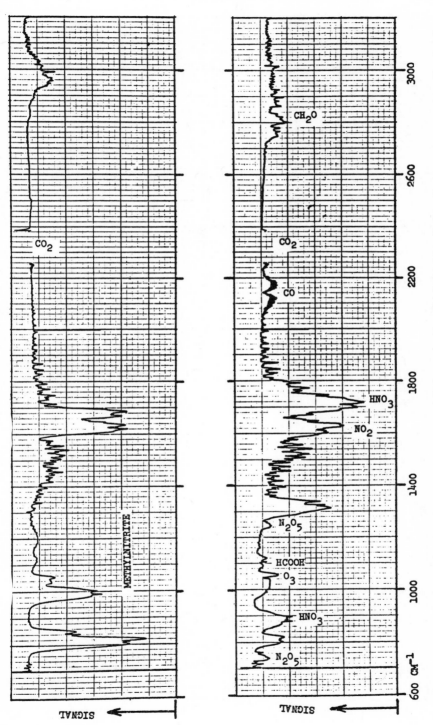

Figure 5. Methyl nitrite photolysis. Top spectrum of methyl nitrite (5 ppm) before irradiation. Lower spectrum after 40 min of irradiation $[k_{d(NO_2)} = 0.3 \ min^{-1}, p = 760 \ torr, 172$-m path length].

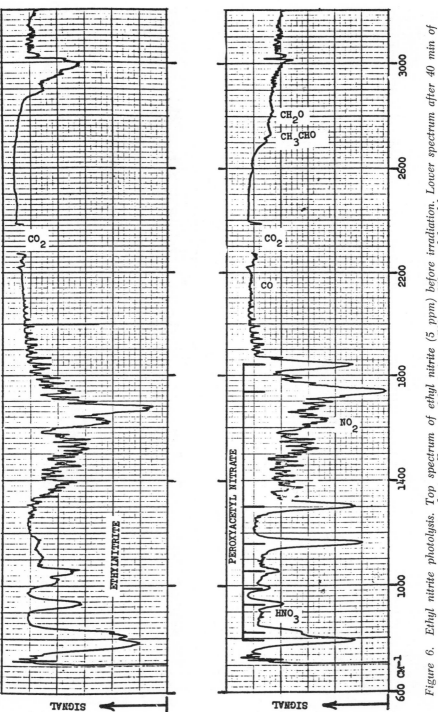

Figure 6. Ethyl nitrite photolysis. Top spectrum of ethyl nitrite (5 ppm) before irradiation. Lower spectrum after 40 min of irradiation [$k_{d(NO_2)} = 0.3\ min^{-1}$, $p = 760\ torr$, 172-m path length].

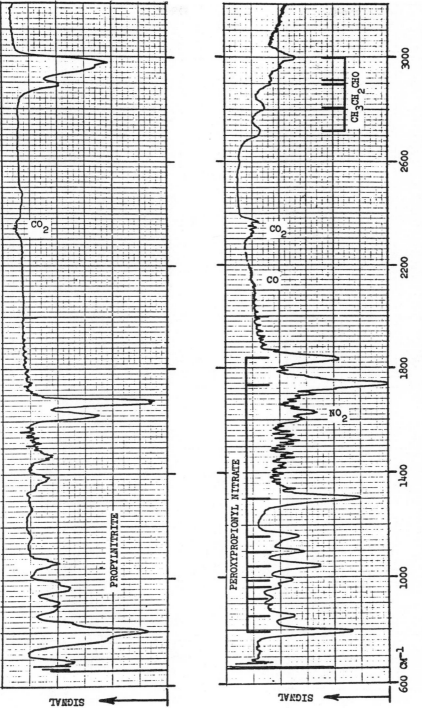

Figure 7. Propyl nitrite photolysis. Top spectrum of propyl nitrite (5 ppm) before irradiation. Lower spectrum after 40 min of irradiation [$k_{d(NO_2)} = 0.3\ min^{-1}$, $p = 760\ torr$, 172-m path length].

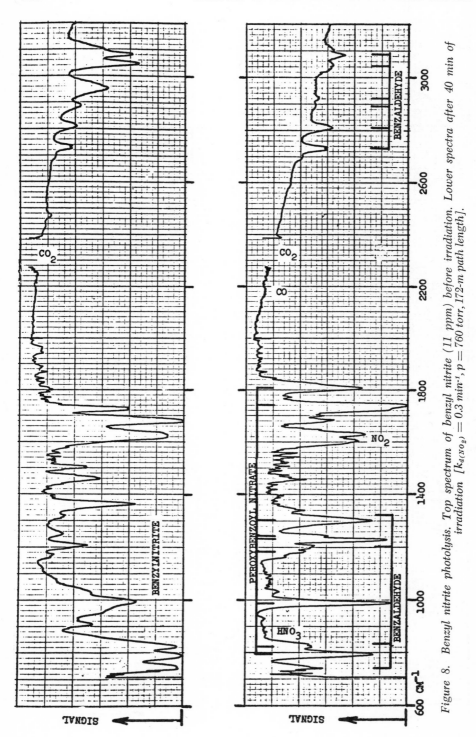

Figure 8.   Benzyl nitrite photolysis. Top spectrum of benzyl nitrite (11 ppm) before irradiation. Lower spectra after 40 min of irradiation $[k_{d(NO_2)} = 0.3\ min^{-1},\ p = 760\ torr,\ 172\text{-}m\ path\ length].$

nitric acid is one of the products, one could postulate it came from PFN dissociation. However, when one looks at the carbon balance of the system (Figure 1) at 60 minutes, all of the carbon can be accounted for as CO, formaldehyde, formic acid, and unreacted methyl nitrite. Thus, there is little latitude for any CO_2 from the PFN dissociation, which reinforces the argument that PFN does not form at any appreciable rate and that the concentration is extremely low, if any is formed at all.

PFN has been identified in this laboratory by its IR spectrum which is similar to that of PAN. It was formed by the gas phase photolysis of chlorine in the presence of formaldehyde, NO_2, and oxygen. The compound was unstable.

$$Cl_2 \overset{h\nu}{\rightarrow} 2\ Cl$$

$$Cl + H_2CO \rightarrow HCl + HCO$$

$$HCO + O_2 \rightarrow H\overset{\overset{\textstyle O}{\textstyle \|}}{C}OO$$

$$H\overset{\overset{\textstyle O}{\textstyle \|}}{C}OO + NO_2 \rightarrow H\overset{\overset{\textstyle O}{\textstyle \|}}{C}OONO_2$$

The concentrations of formyl and peroxyformyl radicals are expected to be much higher in this system than those formed in MeONO photolysis. The addition of NO_2 to the peroxyformyl radical, which itself must be unstable, forms PFN. A thorough investigation has yet to be conducted to determine whether PFN decomposes to nitric acid and CO_2 by the cyclic mechanism described above.

The nitric acid concentration was higher by a factor of 10 in the methyl nitrite system as compared to the other nitrite photolyses (Table 1); 22 percent of the nitrogen from the reacted nitrite was nitric acid in the gas phase. The actual concentrations of nitric acid formed in the photolysis were higher than those measured since substantial amounts were absored onto the cell surface. This fact can explain why the gas phase nitrogen balance accounts for only 35 percent of the nitrogen in the methyl nitrite system. Infrared spectra of the cell evacuated (1.5 torr), plotted in a nonratio plot mode (single beam), showed absorbance due to nitrate on the mirror surface. No attempt was made to quantitate this absorbance nor were attempts made to wash the cell and determine the amount of nitric acid absorbed. In the photooxidation of hydrocarbons in the presence of oxides of nitrogen all of nitrogen, in the system could be accounted for when nitric acid

Table I

Organic Nitrite Photolysis

	MeONO[a]	EtONO[a]	PrONO[b]	BzONO[a]
Initial Conc. ppm	5.5	5.0	5.2	6.4
Amount left (ppm) After 60 min. Irradiation	0.21	0.16	0.10	trace[c]
Percent Products (*)				
HNO_3	22	2	2.5	1.3
NO_2	11	6	8	25
PAN type	0	64	62	59
Formic Acid	12	4	trace[c]	trace[c]
CO	49	6	6	5
Aldehyde	36[d]	22[e]	35[f]	30[g]

Table I (CONTINUED)

(a) $k_{d(NO_2)} \simeq 0.3$ min^{-1}

(b) $k_{d(NO_2)} \simeq 0.6$ min^{-1}, values are for 30 min. irradiation

(c) less than 0.04 ppm compound

(d) Formaldehyde

(e) 15 percent acetaldehyde, 7 percent formaldehyde

(f) 33 percent propionaldehyde, 2 percent formaldehyde

(g) Benzaldehyde, trace of formaldehyde observed < 0.04 ppm

(*) Percent products were calculated as concentration of products ppm divided by initial concentrations of organic nitrite ppm times one hundred. The nitrogen containing material being HNO_3, NO_2, PAN-type compound and unreacted nitrite; carbon containing material being PAN-type compound, formic acid, CO, aldehyde, and unreacted nitrite.

on the reactor surface was considered (17). When methyl
nitrite was photolyzed at 30 percent relative humidity, the
amount of nitric acid in gas phase decrease. This is due to a
two fold effect water has on the system. First water vapor
enhances the formation of nitric acid by (reaction 14).
Second the nitric acid formed in the gas phase coalesces with
water molecules and becomes absorbed on the reactor surface.
The relative amounts of the carbon containing products also
changed when the relative humidity was varied (Table III).
This phenomenum is indicative of either a change in importance
of rates in the overall reaction scheme and/or a change in the
reaction mechanisms in the presence of appreciable amounts of
water.

Wiebe (18) observed N_2O as a product in the photolysis of
methyl nitrite at torr level concentrations. They postulated
its formation by this reaction:

$$2 \; HNO \rightarrow H_2O + N_2O$$

There was no indication from our infrared spectral data that
any N_2O was formed in any of the nitrite systems studied.

Methyl nitrate did not form when low concentrations of
methyl nitrite were photolyzed in air or oxygen. However,
methyl nitrate was formed in the photolysis of methyl nitrite
in nitrogen, and in an oxygen system at initial nitrite concentrations higher than 8 ppm. Methyl nitrate, which is not
readily photodissociated by the uv light from the lamps used,
has a long half life in our system and can be detected at low
concentrations.

Nitric acid is the sink for nitrogen in the case of methyl
nitrite photolysis, whereas PAN type compounds are sinks in the
other systems. Long path infrared studies of the ambient atmosphere during high photochemical activity have shown PAN to
be the major nitrogen-containing product in a well reacted air
mass. Formic acid was the only other major carbon hydrogen containing material in this well reacted air mass (1).

With the photolysis of ethyl nitrite significant C-C bond
fracture takes place as observed in the products: 7 percent
formaldehyde, 6 percent CO, and 4 percent formic acid. In the
case of propyl nitrite there was 6 percent CO, 2 percent
formaldehyde and only a trace of formic acid. In the benzyl
nitrite system only traces of one carbon-component materials
were detected.

Conclusion

The photolyses of various alkyl and benzyl nitrites gave
similar reaction products, with the exception of methyl nitrite
where no PAN-type compounds were formed.

In the photolysis of methyl nitrite, nitric acid was the

Table II

Carbon and Nitrogen Balance of Organic Nitrite Photolysis

Compound	Percent Accountable	
Photolyzed	Carbon	Nitrogen
MeONO	100	35
EtONO	90	75
PrONO	99	75
BzONO	90	84

Table III

Gas Phase Products vs Relative Humidity

of Methyl Nitrite Photolysis

	Percent Product	
Product	< 1 percent RH	30 percent RH
HNO_3	22	10
Formic Acid	12	31
CH_2O	36	31
CO	49	33

major nitrogen-containing product. No PFN was observed nor did the carbon balance allow for CO_2 formation by the cyclic decomposition of PFN. Amounts of ozone and N_2O_5 were much higher in this system. A change in percent relative humidity had a marked effect on percentages of products. This indicates that some change in the rates of reaction or in the mechanisms occurred.

    The major stable product in the photolysis of nitrites other than methyl nitrite was a PAN-type compound. In the ethyl and propyl nitrite photolyses there was some carbon-carbon bond breakage, whereas very little rupture of the ring occurred in benzyl nitrite photolysis. The carbon balances for all systems were good and the nitrogen balances for the ethyl, propyl, and benzyl nitrite system were better than the methyl nitrite photolysis. This is undoubtedly due to the fact that the major nitrogen-containing materials were PAN-type compounds, which stayed in the gas phase. The benzyl nitrite system had the best nitrogen balance since it had the least amount of nitric acid formed.

Literature Cited

1. Hanst, P.L., Wilson, W.E., Patterson, R.K., Gay, Jr., B.W., Cheney, L.W., Burton, S.C., Environmental Protection Agency Report, EPA-650/4-75-006, Feb. 1975.

2. Taylor, O.C., J. Air. Poll. Control Assoc., 19, 347 (1969).

3. Lonneman, W.A., Private communication of unpublished data at field studies at New York, Los Angeles, St. Louis, (1975).

4. Darley, E.F., Kettner, K.A., Stephens, E.R., Anal. Chem., 35: 589-591 (1963).

5. Heuss, J.M., Glasson, W.A., Environ. Sci. & Technol., 2, 1109 (1968).

6. Demerjian, K.L., Kerr, J.A., Calvert, J.G., "The Mechanism of Photochemical Smog Formation", Adv. Enviro. Sci. & Technol., 4, 1 (1974).

7. Stephens, E.R., Hanst, P.L., Doerr, R.C., Scott, W.E., Inc. Eng. Chem. 48, 1498 (1956).

8. Stephens, E.R., Scott, W.E., Hanst, P.L., Doerr, R.C., J. Air Poll. Control Assoc., 6, 159 (1956).

9. Tuesday, C.S., in "Chemical Reactions in the Lower and Upper Atmosphere", Cadle, R.D., Editor, Interscience, N.Y., p. 1-49, (1961).

10. Stephens, E.R., Burleson, F.R., Cardiff, E.A., J. Air. Poll. Control Assoc., 15 87 (1965).

11. Gay, Jr., B.W., Noonan, R.C., Bufalini, J.J., Hanst, P.L., Photochemical Synthesis of Peroxyacyl Nitrates in the Gas Phase via Chlorine-Aldehyde Reaction, submitted for publication to Env. Sci. and Technol., (1975).

12. Hanst, P.L., Adv. in Environ. Sci. & Technol., Edited by Pitts and Metcalf, Published by John Wiley and Sons, Inc. (1971).

13. Hanst, P.L., Lefohn, A.S., Gay, Jr., B.W., Applied Spectroscopy, Vol. 27, #3, 188 (1973).

14. Black, A.P., Babers, F.H., Organic Syntheses, 2, 412 (1943).

15. Chretien, A., Comp. Rend., 200, 746 (1945).

16. Hecht, T.A., Seinfeld, J.H., Dodge, M.C., Environ. Sci. & Technol., 4, 327 (1974).

17. Gay, Jr., B.W., Bufalini, J.J., Environ. Sci. and Technol., 5, 423 (1971).

18. Wiebe, H.A., Villa, A., Hellman, T.M., Heicklen, J., J. Am. Chem. Soc., 95, 7 (1973).

13

Fates and Levels of Ambient Halocarbons

DANIEL LILLIAN[1], HANWANT BIR SINGH[2], ALAN APPLEBY, LEON LOBBAN, ROBERT ARNTS, RALPH GUMPERT, ROBERT HAGUE, JOHN TOOMEY, JOHN KAZAZIS, MARK ANTELL, DAVID HANSEN, and BARRY SCOTT

Department of Environmental Science, Rutgers University, New Brunswick, N. J. 08903

There is much recent concern over the behavior and effects of halocarbons in the environment. Ambient CCl_3F and CCl_2F_2 (1,2) and, by analogy, other tropospherically stable compounds are suspect as precursors of stratospheric ozone-destroying chlorine atoms. Vinyl chloride has been linked to industrial angiosarcoma and is possibly mutagenic (3). Chloroethylenes will react to form significant quantities of highly toxic phosgene and acetyl chlorides under simulated tropospheric conditions (4,5). Chloroform and CCl_4 in Mississippi drinking water have been associated with an elevated cancer risk (6,7). Concurrent to these findings, the research effort in ambient halocarbon measurement has justifiably increased significantly (8-16). The many industrial and domestic uses of halogenated hydrocarbons and their rather large production figures (12 billion pounds in 1974 for the U.S. alone) (17) suggest that this accelerated research effort will lead to a continual increase in the number of halocarbons routinely measured in the environment. Indeed, mass spectrometric analysis conducted on cryogenically concentrated air samples several years ago indicated the presence of a wide variety of halogenated compounds (18).

Because of the large number of atmospheric halogenated hydrocarbons of potential interest, a selective study of a carefully chosen group representing a wide spectrum of chemical reactivities and emission patterns was clearly desirable. This approach would not only provide information of immediate relevance but also contribute to a data base for future reference as the number of halocarbons of environmental interest proliferates.

A comprehensive study of the atmospheric chemistry and

(1) Currently with the U.S. Army Industrial Hygiene Agency, Edgewood Arsenal, Maryland.
(2) Currently with the Stanford Research Institute, Menlo Park, California.

ambient concentrations of halogenated hydrocarbons has been
underway in our laboratory since July 1972. Presented in this
first of a series of reports are ambient data obtained at several
locations for CCl_3F, CH_3I, CCl_4, C_2Cl_4, CH_3CCl_3 and $CHCl_3$. To
address the possible fate of these compounds representative concentration-time profiles illustrating their behavior under experimentally simulated tropospheric conditions are presented.

Experimental

Ambient halocarbons were measured on site using a mobile
laboratory, equipped with a Fisher-Victoreen gas chromatograph
and a special coulometric gas chromatograph. The analytical procedures have been discussed in detail (6). The minimum detectable concentrations for CCl_3F, CH_3I, CCl_4, C_2Cl_4, CH_3CCl_3 and
$CHCl_3$ based on a 5 ml injection were respectively .002, .002,
.01, .04, .02 and .006 ppb.

The "smog chamber" experiments were conducted using plastic
film bags or a 72 liter Pyrex glass reactor. Simulated tropospheric sunlight for bag irradiations was provided by a bank of
mixed sun, black light and black light-blue lamps ($k_1=0.4min^{-1}$).
The 72 liter reactor was irradiated in an air conditioned chamber
fitted with slim-line black lights (42T6) ($k_1=0.3min^{-1}$). A non-
leaded fuel (60% paraffin, 13% olefin, 27% aromatic) was used to
simulate tropospheric hydrocarbon reactivity (19). Ozone and
oxides of nitrogen were measured by gas-phase chemiluminescence.
Stratospheric simulations were conducted using an Ace Glass 1
liter photochemical reactor equipped with a Hanovia 450 watt
quartz lamp ($\lambda>2200$Å). Gas blending procedures and other experimental techniques have been discussed (19,20). During extended
irradiations of stable halocarbons NO_2 was replenished every 24
hours.

Results

Figure 1 shows the respective average ambient concentrations
of CCl_3F, CH_3I, CCl_4, C_2Cl_4, CH_3CCl_3 and $CHCl_3$ measured at the
locations and times indicated. Also shown are the concentration
ranges for each substance expressed as maximum and minimum levels
detected. Figure 2 shows the concentration-time profiles for
CCl_3F, CCl_4 and CH_3CCl_3 obtained when the indicated halocarbon
was irradiated with simulated tropospheric sunlight in the presence of a reactive hydrocarbon mixture and nitrogen dioxide.
Figure 3 shows the decay of CCl_4 and CCl_3F when irradiated with
U.V. Figure 4 illustrates the effect of simulated tropospheric
sunlight on C_2Cl_4 in the presence of NO_2 and reactive hydrocarbon. The photochemical decay of CH_3I in air using the same
illumination is shown in Figure 5.

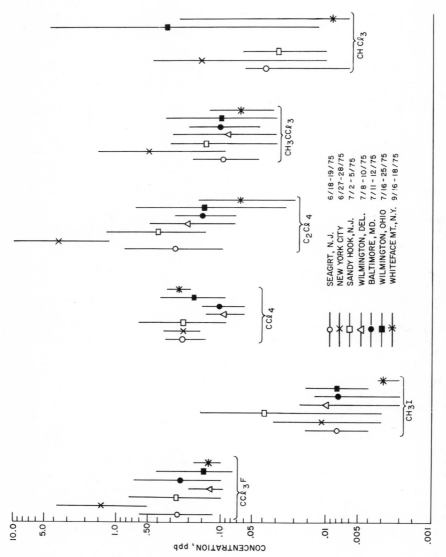

Figure 1.   Halocarbon values at various locations averages maxima and minima

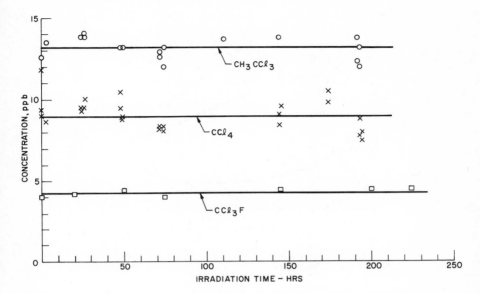

Figure 2. *CH₃CCl₃, CCl₄, or CCl₃F irradiated in ultra zero air in Mylar bags containing 1 ppm hydrocarbon, 0.5 ppm NO₂, and 50% relative humidity*

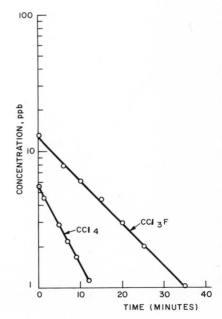

Figure 3. *UV (λ > 2200 A) halocarbon decay in ultra zero air*

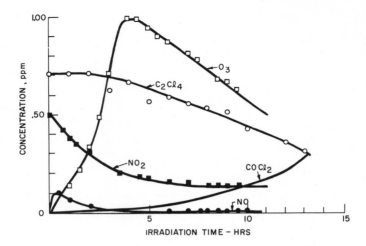

Figure 4. C_2Cl_4 irradiated in ultra zero air in a Mylar bag containing 1 ppm hydrocarbon, 0.5 ppm NO_2, and 50% relative humidity

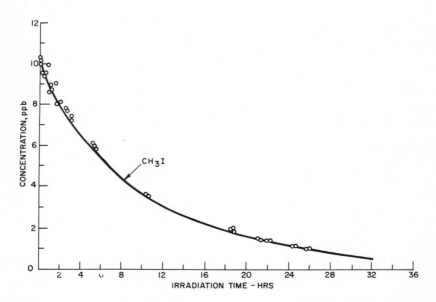

Figure 5. CH_3I irradiated in ultra zero air in a Mylar bag containing 50% relative humidity

Discussion

    Figure 1 clearly shows that the halocarbons measured are
extremely variable in concentration at all the locations studied,
and from one location to another. Some general observations can
however be made. The highest values of all compounds were observed at New York City, except for CH_3I where the New York City
value was the second highest. Whiteface Mountain had been expected to exhibit the lowest halocarbon levels because of its
remoteness from highly industrialized or populated areas. However average CCl_3F and CCl_4 concentrations were observed to be
lower in other areas. CCl_4 was always measurable, and exhibited
the least variability. The minimum value observed by us agrees
well with that reported as background concentration by Lovelock
et al. and Wilkness et al. (about 0.07 ppb) (10,13). CCl_3F was
similarly always measurable and our minimum concentrations were
reasonably close to reported background values (9,10,13) of about
0.05 to 0.09 ppb.
    CCl_4 and CH_3CCl_3 are, like CCl_3F tropospherically inert, as
is shown in Figure 2. Accordingly, these compounds would serve
as sources of stratospheric chlorine atoms, similar to CCl_3F and
CCl_2F_2 and as Figure 1 shows, in roughly equivalent quantities.
Figure 3 demonstrates the stratospheric reactivity of CCl_3F and
CCl_4.
    Perchloroethylene, used primarily in dry-cleaning operations,
is probably not a significant direct source of stratospheric
chlorine atoms because of its tropospheric reactivity (Fig. 4).
Its environmental significance lies in its possible tropospheric
conversion to phosgene. Bearing in mind the need for caution in
extrapolating smog chamber data to the environment, it is noteworthy that recent laboratory experiments (4,5,21) have demonstrated as much as 84% by weight conversion of C_2Cl_4 to highly
toxic phosgene. (1974 proposed TLV is 50 ppb, not to be exceeded). There is evidence also that another tropospheric
degradation product of C_2Cl_4 is CCl_4 (4) which may account for a
significant portion of the CCl_4 tropospheric budget.
    CH_3I was not always measurable and, in common with all the
compounds reported here, undetectable concentrations were not
included in the averages. The detection frequency for CH_3I and
other compounds are discussed elsewhere (22); it is noteworthy
here however that CH_3I was generally most frequently measurable
in close proximity to the ocean, consistent with the suggestion
of Lovelock et al. (10) that its origins are in the ocean and
are biological. The values we obtain are also not inconsistent
with their mean aerial concentration of 0.0012±0.01 ppb. The
highest average of CH_3I concentration was at Sandy Hook, on the
New Jersey shore. The lower value at nearby Seagirt is probably
due to a prevailing offshore breeze during the sampling period.
    CH_3I is very reactive, and is expected to undergo rapid decay
in the troposphere (Fig. 5). Its decreasing frequency of

observation away from the oceans and the sporadic nature of its
detectability is presumably a manifestation of this reactivity.

Literature Cited

1. Molina, M.J., and Rowland, F.S., Nature, (1974), 249, 810.
2. Cicerone, R.J., Stolarski, R.S., and Walters, S., Science,
    (1974), 185, 1165.
3. Chemical and Engineering News, (1974), 52, 6.
4. Singh, H.B., Lillian, D., Appleby, A., and Lobban, L.A.,
    Environmental Letters, submitted March 1975.
5. Lillian, D., Singh, H.B., Appleby, A., and Lobban, L.A.,
    J. Air Poll. Control Assoc., accepted for publication,
    February 1975.
6. Chemical and Engineering News, (1974), 52, 5.
7. Chemical and Engineering News, (1974), 52, 44.
8. Lovelock, J.E., Nature, (1971), 230, 379.
9. Hester, N.E., Stephens, E.P., and Taylor, O.C., J. Air Poll.
    Control Assoc., (1974), 24, 591.
10. Lovelock, J.E., Maggs, R.J., and Wade, R.J., Nature, (1973),
    241, 194.
11. Lillian, D., and Singh, H.B., Anal. Chem., (1974), 46, 1060.
12. Lovelock, J.E., Atm. Env., (1972), 6, 917.
13. Wilkness, P.E., Lamontagne, R.A., Larson, R.E., and
    Swinnerton, J.W., Nature Phys. Sci., (1973), 245, 45.
14. Murray, A.J., and Riley, J.P., Nature, (1973), 242, 37.
15. Chih-Wu-Su, and Goldberg, E.D., Nature, (1973), 2245, 27.
16. Simmonds, P.G., Kerrin, S.L., Lovelock, J.E., and Shair, F.H.,
    Atmos. Env., (1974), 8, 209.
17. United States International Trade Commission Preliminary
    Report, February 5, 1975.
18. Weaver, E.R., Hughes, E.E., Gunther, S.M., Schuhmann, S.,
    Redfearn, N.T., Gorden, R., J. Res. Nat. Bur. Stand., (1957),
    59, 383.
19. Lillian, D., and Hansen, D., "Aerosol formation from the
    photo-oxidation of non-leaded fuels," 164th National A.C.S.
    meeting, Division of Water and Water Chemistry, N.Y.
    August 27 - September 1 (1972).
20. Ripperton, L.A., and Lillian D., J. Air Poll. Control Assoc.,
    (1971), 21, 679.
21. Singh, H.B., Lillian, D., and Appleby, A., Anal. Chem.,
    accepted for publication, January 1975.
22. Lillian, D., Singh, H.B., Appleby, A., Lobban, L.A., Arnts,
    R., Gumpert, R., Hague, R., Toomey, J., Kazazis, J., Antell,
    M., Hansen, D., and Scott, B., Env. Sci. and Tech., sub-
    mitted November 1974.

The Fate of Nitrogen Oxides in Urban Atmospheres

CHESTER W. SPICER,* JAMES L. GEMMA, DARRELL W. JOSEPH, and ARTHUR LEVY

Battelle, Columbus Laboratories, Columbus, Ohio 43201

Nitrogen oxides enter the atmosphere from a variety of sources, principally from automotive exhausts and power plant combustion. Some nitrogen species, such as ammonia resulting from the natural decomposition of organic material, are also an eventual source of nitrogen oxides. Most of the nitrogen oxides are converted into nitrogen dioxide which, in turn, reacts with other species in the atmosphere producing nitrates and nitrites in both gaseous and particulate forms. However, the ultimate disposition of the nitrogen oxides is unknown. The problem of uncovering the fate of nitrogen oxides in the atmosphere is complicated by the continuous movement of air masses and the continual input of material into the atmosphere. The current program was designed to measure a multitude of chemical and meteorological parameters in two urban atmospheres in order to determine the distribution and disposition of nitrogen compounds in the atmosphere and the factors which influence the nitrogen oxides removal processes.

The program described here consisted of three distinct phases: analytical methodology development, field sampling, and data analysis. The analytical development phase of the program has involved testing state-of-the-art techniques and in some instances development and validation of novel procedures for determining the following species during the field sampling phase of the program: NO, NO_2, O_3, CH_3ONO_2, $C_2H_5ONO_2$, PAN, NH_3, HNO_3, mass loading, particulate NO_3^-, particulate NO_2^-, particulate NH_4^+, and total C, H, and N in the particulate phase. Temperature, relative humidity, wind speed, wind direction, and solar intensity were also monitored. Some rainfall and dustfall samples were collected and analyzed for nitrogen species.

The field sampling phase of the program was carried out in the summer of 1973, in St. Louis, Missouri, and West Covina,

---

*To whom correspondence should be addressed.

California (located 25 miles east of downtown Los Angeles).

Experimental

The analytical methods employed in this investigation are listed in Table I. Both NO and NO_2 were monitored by chemiluminescence techniques. We have previously reported[1] on interference to the chemiluminescent determination of $\overline{NO_2}$ by a variety of other gaseous nitrogen species including PAN, HNO_3, and alkyl nitrates. Ozone was monitored by chemiluminescence methods. A dual catalytic-converter chemiluminscent instrument was used to monitor ammonia. Gas phase organic nitrates were determined by electron capture detection gas chromatography. Two novel methods for HNO_3 determination were developed. One, a modified colorimetric procedure[2], yields integrated HNO_3 values, while the other, based on $\overline{coulometry}$[3], gives a continuous readout of atmospheric nitric acid.

Table I.   Analytical Methods

Measurement	Technique
NO, NO_2, NO_x	Chemiluminescence
O_3	Chemiluminescence
NH_3	Chemiluminescence
PAN	Electron Capture Gas Chromatography
Alkyl Nitrates	Electron Capture Gas Chromatography
HNO_3	Continuous Coulometric
HNO_3	Integrated Colorimetric
NH_4^+	
NO_3^-	Hi Vol. Sampling and Standard Analysis
NO_2^-	
Total C, H, N	
Wind Speed	
Wind Direction	
Relative Humidity	MRI, INC. Weather Station
Temperature	
Sunlight Intensity	Pyrohelimoter

Aerosol samples were collected on high-purity quartz-

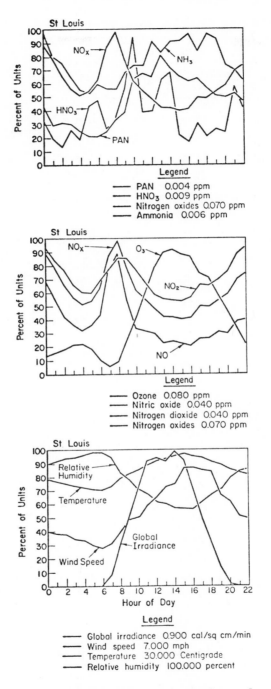

Figure 1. Average diurnal air quality and meteorological profile, St. Louis

Table II.   Summary of

Date	Day	Weather Conditions General[a]	Temp, C	RH%	Aerosol Mass Loading $\mu g/m^3$	O_3, ppm	NO, ppm	NO_x, ppm
7-18	W	S	28	75	114.1	0.049	0.012	0.061
7-19	Th	R, S	27	92	80.7	0.038	0.016	0.061
7-20	F	S	29	79	51.5	0.027	0.013	0.047
7-22	Sun	S	27	82	32.6	0.072	0.019	0.037
7-23	M	S, R	26	84	42.8	0.041	0.005	0.034
7-24	T	S	27	86	37.0	0.037	0.011	0.048
7-25	W	S	27	80	53.9	0.044	0.021	0.056
7-26	Th	R, C	26	69	33.4	0.032	0.013	0.035
7-27	F	C	27	67	50.9	0.028	0.013	0.038
7-30	M	R	25	76	68.2	0.023	0.026	0.057
7-31	T	PC, R	24	72	39.2	0.032	0.015	0.034
8-1	W	C	21	72	41.9	0.022	0.016	0.028
8-2	Th	C	22	66	43.5	0.022	0.014	0.034
8-3	F	S	23	61	53.7	0.034	0.011	0.035
8-4	Sat	S	25	68	79.9	0.052	0.007	0.033
8-5	Sun	S	26	76	63.0	0.045	0.009	0.039
8-6	M	PC	25	72	74.9	0.042	0.011	0.038
8-7	T	S	27	84	61.4	0.038	0.012	0.037
8-8	W	S	30	82	51.5	0.023	0.014	0.036
8-9	Th	R	24	95	47.9	0.004	0.024	0.051
8-10	F	S, R	24	82	66.4	0.041	0.015	0.041
8-12	Sun	S, R	26	85	40.6	0.037	0.026	0.047
8-13	M	R	23	93	63.3	0.016	0.026	0.054
8-14	T	R, S	21	82	42.0	0.035	0.013	0.029
8-15	W	PC	23	76	90.5	0.041	0.034	0.062
8-16	Th	R, S	23	83	79.6	0.029	0.016	0.044

(a) S = Sunny, R = Rain, C = Clear, PC = Partly Cloudy, Cl = Cloudy.
(b) Continuous Coulometric.

Table III.   Summary of St. Louis Hydrocarbon Data[a]

Date Sample Number[b] Component	7-18-73 1	2	3	4	7-19-73 1	2	3	4
Methane	2420	2541	2390	2053	2057	2326	2271	2235
CO	1674	2008	1266	578	1405	1746	2134	1634
C_2H_2	21	24	18	5	17	26	38	27
C_2H_4	29	29	17	7	30	28	40	26
Olefins	80	88	54	34	77	92	103	73
Aromatics	—	—	—	—	—	—	—	—
Nonmethane Hydrocarbons	570	454	452	140	295	435	555	424/
C_2H_4/C_2H_2	1.38	1.22	0.93	1.42	1.79	1.06	1.03	0.96

(a) Courtesy of the U.S. Environmental Protection Agency.

Air Quality Data, St. Louis

	Pollutants								
23-Hour Average				1-Hour Maximum					
NH_3, ppm	PAN, ppm	HNO_3[b], ppm	HNO_3[c], ppm	O_3, ppm	NO, ppm	NO_x, ppm	NH_3, ppm	PAN, ppm	HNO_3, ppm
0.010	0.001	0.012	0.007	0.110	0.043	0.114	0.014	0.003	0.044
0.008	0.002	0.005	0.004	0.107	0.061	0.121	0.011	0.007	0.022
0.008	0.002	0.002	0.000	0.083	0.034	0.070	0.011	0.006	0.012
0.002	0.001	0.001	0.007	0.146	0.026	0.050	0.016	0.003	0.003
0.004	0.001	0.006	0.009	0.086	0.020	0.067	--	0.002	0.055
0.006	0.001	0.004	0.011	0.113	0.039	0.081	0.011	0.004	0.016
0.005	0.001	0.002	0.002	0.128	0.070	0.108	0.009	0.004	0.017
0.004	0.001	0.000	0.004	0.066	0.024	0.060	0.005	0.001	0.000
0.004	0.001	0.000	0.004	0.067	0.042	0.069	0.006	0.004	0.000
--	--	0.001	0.006	0.050	0.054	0.089	--	--	0.006
0.004	0.001	--	0.002	0.068	0.047	0.084	0.005	0.002	--
0.002	0.001	--	0.003	0.033	0.030	0.049	0.003	0.002	--
0.003	0.001	--	0.005	0.051	0.033	0.061	0.005	0.002	--
0.002	0.002	0.000	0.004	0.087	0.037	0.077	0.004	0.004	0.000
0.003	0.003	0.004	0.008	0.124	0.051	0.096	0.006	0.006	0.015
0.003	0.002	0.003	0.001	0.092	0.034	0.091	0.006	0.003	0.018
0.002	0.001	0.002	0.002	0.078	0.035	0.075	0.004	0.003	0.029
0.002	0.002	0.005	0.004	0.076	0.030	0.067	0.005	0.004	0.020
0.003	0.002	0.003	0.004	0.043	0.031	0.059	0.006	0.004	0.021
0.003	0.002	0.001	0.000	0.011	0.060	0.089	0.007	0.004	0.011
0.002	0.005	0.004	0.002	0.163	0.083	0.106	0.004	0.019	0.024
--	0.003	0.001	0.007	0.138	0.101	0.133	--	0.006	0.005
0.005	0.002	0.008	0.000	0.058	0.051	0.094	0.008	0.004	0.042
0.002	0.002	0.000	0.000	0.063	0.021	0.059	0.004	0.004	0.004
0.003	0.003	0.009	0.001	0.096	0.103	0.147	0.009	0.009	0.080
0.002	0.003	0.003	0.000	0.065	0.042	0.078	0.005	0.008	0.017

(c) Integrated Colorimetric.

(Hydrocarbon values in ppbC; carbon monoxide values in ppb.)

7-20-73				7-23-73				7-24-73			
1	2	3	4	1	2	3	4	1	2	3	4
2310	2418	1984	1570	2445	2235	2094	2079	2107	2129	2285	2117
1536	2047	594	390	882	1018	817	528	1010	1423	1113	930
27	23	12	10	12	14	4	4	15	21	13	12
24	30	15	11	20	18	8	9	20	26	17	15
75	100	60	39	73	53	35	36	66	74	62	41
140	273	166	109	129	353	220	217	155	198	78	46
513	818	530	376	448	657	448	420	486	579	387	284
0.88	1.32	1.22	1.08	1.62	1.31	1.91	2.09	1.30	1.25	1.31	1.26

(b) Sample No. 1 = 6-8 a.m.; 2 = 8-10 a.m.; 3 = 10 a.m.-12 p.m.; 4 = 12-2 p.m

fiber filters by two high volume samplers. Conventional wet
chemistry procedures were employed for NO_3^- and NO_2^- determination. Ammonium ion was determined, after conversion to ammonia,
by an ammonia gas-sensing electrode. A combustion technique
was used for total C, H, N analysis.

Results

Monitoring of chemical and meteorological variables was
carried out for 5 weeks in St. Louis, Missouri, and 5 weeks in
West Covina, California. Sampling in both cities was conducted
on a 23-hour-per-day basis, from 11:30 p.m. to 10:30 p.m. A
summary of the air-quality data collected in St. Louis is shown
in Table II. The gas-phase data are presented as both 23-hour
averages and as maximum 1-hour averages. A summary of St. Louis
hydrocarbon and CO data taken by the Environmental Protection
Agency for several days simultaneously with our measurements is
presented in Table III. Figure 1 displays the profiles of the
average diurnal air quality and meteorology during the St. Louis
field program. These profiles are composites over the entire
5-week sampling period. The averaged aerosol results from our
St. Louis study are given in Table IV for two particle-size
fractions.

Table IV.   Aerosol Analysis

| | Weight Percent | | | | | |
	NH_4^+	NO_2^-	NO_3^-	C	H	N
St. Louis, Missouri						
Average Total Aerosol Composition	4.4	–	0.63	19.0	3.6	4.6
Large Particle (>2.5 μm) Composition	0.55	0.001	2.65	14.6	1.8	1.5
West Covina, California						
Average Total Aerosol Composition	4.7	–	1.7	19.4	3.8	5.3
Large Particle (>2.5 μm) Composition	0.63	0.001	4.8	12.6	1.8	2.2

A summary of the air-quality data for the 29 sampling days
in West Covina, California, is shown in Table V. The gas-phase
data are tabulated as both daily (23-hour) averages and maximum
1-hour averages. Figure 2 profiles the average diurnal air
quality and meteorology during the field sampling in West Covina.
The average aerosol results from our West Covina sampling are

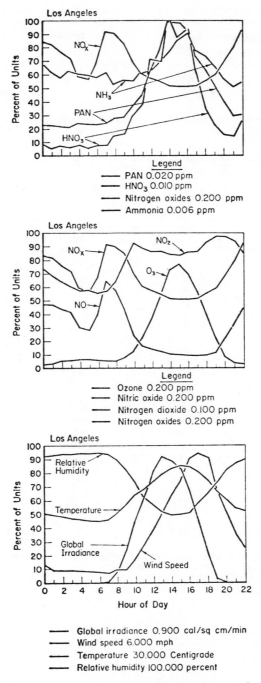

Figure 2. Average diurnal air quality and meteorological profile, West Covina

Table V.   Summary of Air

Date	Day	Weather Conditions			Aerosol Mass Loading $\mu g/m^3$	O_3, ppm	NO, ppm	24 NO$_x$, ppm
		General(a)	Temp. C	RH%				
8-24	F	S	21	89	115.8	0.054	0.071	0.180
8-26	Sun	C	18	93	54.9	0.035	0.046	0.081
8-27	M	S	20	89	65.1	0.039	0.029	0.103
8-28	T	PC	19	--	81.7	0.050	0.061	0.152
8-29	W	PC	20	--	98.9	0.060	0.060	0.171
8-30	Th	PC	19	--	96.1	0.044	0.020	0.120
8-31	F	C	18	--	87.0	0.043	0.021	0.113
9-3	M	S	18	--	64.5	0.042	0.007	0.062
9-4	T	Cl	17	--	79.7	0.021	0.013	0.082
9-5	W	PC,S	18	--	77.6	0.048	0.020	0.090
9-6	Th	S	20	66	118.1	0.083	0.080	0.204
9-7	F	S	18	87	135.8	0.047	0.114	0.239
9-8	Sat	Cl,S	18	79	76.6	0.055	0.045	0.093
9-9	Sun	Cl,S	19	83	63.6	0.036	0.055	0.105
9-10	M	Cl	17	90	64.7	0.017	0.056	0.122
9-11	T	Cl,S	18	79	81.8	0.041	0.025	0.096
9-12	W	Cl,S	17	88	105.5	0.031	0.058	0.151
9-13	Th	Cl,S	17	85	94.1	0.048	0.020	0.071
9-14	F	Cl,S	17	80	123.7	0.064	0.013	0.075
9-17	M	--	16	77	127.4	0.070	0.020	0.109
9-18	T	S	16	82	145.8	0.058	0.046	0.152
9-19	W	PC	17	82	137.3	0.052	0.044	0.140
9-20	Th	--	16	85	116.5	0.048	0.041	0.112
9-21	F	--	18	76	122.7	0.050	--	--
9-24	M	C	16	71	73.1	0.032	0.121	0.163
9-25	T	S	19	83	106.3	0.051	0.037	0.103
9-26	W	S	22	50	198.2	0.040	0.177	0.252
9-27	Th	S	23	37	123.7	0.031	0.147	0.221
9-28	F	C	23	44	105.6	0.052	0.175	0.256

(a) S = Sunny, R = Rain, C = Clear, PC = Partly Cloudy, Cl = Cloudy.

(b) Continuous Coulometric.

(c) Integrated Colorimetric.

Quality Data, West Covina

			Pollutants						
Hour Average				1-Hour Maximum					
NH_3, ppm	PAN, ppm	HNO_3[b], ppm	HNO_3[c], ppm	O_3, ppm	NO, ppm	NO_x, ppm	NH_3, ppm	PAN, ppm	HNO_3, ppm
0.000	0.003	0.009	0.010	0.173	0.202	0.296	0.002	0.004	0.029
--	--	0.001	0.002	0.068	0.089	0.166	--	--	0.003
0.008	--	0.000	0.000	0.142	0.065	0.143	0.013	--	0.000
0.007	--	0.002	0.003	0.210	0.274	0.360	0.013	--	0.011
0.005	--	0.003	0.000	0.264	0.205	0.272	0.008	--	0.015
0.005	--	0.002	0.000	0.171	0.045	0.173	0.007	--	0.009
0.004	0.001	0.000	0.001	0.157	0.051	0.161	0.006	0.003	0.000
0.002	0.002	0.002	0.005	0.132	0.042	0.099	0.003	0.006	0.015
0.002	0.003	0.001	--	0.088	0.032	0.125	0.003	0.006	0.005
0.002	0.007	0.000	0.000	0.171	0.052	0.127	0.003	0.015	0.009
0.002	0.013	0.012	0.018	0.271	0.324	0.445	0.005	0.036	0.040
0.004	0.008	0.011	0.003	0.180	0.346	0.487	0.005	0.027	0.022
0.002	0.007	0.004	0.009	0.147	0.065	0.143	0.003	0.017	0.015
0.002	0.005	0.001	--	0.103	0.068	0.127	0.003	0.010	0.005
0.002	0.003	0.002	--	0.040	0.090	0.160	0.003	0.008	0.013
0.002	0.007	0.002	0.006	0.184	0.048	0.129	0.003	0.023	0.010
0.002	0.006	0.002	0.000	0.151	0.176	0.253	0.003	0.027	0.010
--	0.010	0.001	0.010	0.154	0.062	0.121	--	0.025	0.007
--	0.018	0.005	0.026	0.209	0.022	0.126	--	0.040	0.024
--	0.018	0.007	0.016	0.234	0.071	0.135	--	0.042	0.031
--	0.020	0.010	0.018	0.213	0.092	0.227	--	0.046	0.034
--	0.016	0.004	0.019	0.227	0.133	0.202	--	0.044	0.015
--	0.012	0.004	0.008	0.210	0.110	0.199	--	0.040	0.016
--	0.010	--	0.000	0.170	--	---	--	0.025	--
--	0.007	--	0.007	0.132	0.439	0.409	--	0.019	--
--	0.012	0.002	0.011	0.176	0.277	0.403	--	0.029	0.010
--	0.008	0.000	0.007	0.150	0.407	0.470	--	0.013	0.000
--	0.006	0.000	0.009	0.152	0.453	0.706	--	0.008	0.000
--	0.009	0.002	0.014	0.173	0.557	0.691	--	0.025	0.009

shown in Table IV.

Discussion

A detailed discussion of the behavior of the gaseous and aerosol species monitored in St. Louis and West Covina has been presented elsewhere[2] and will not be dwelt on here. Instead this discussion is directed toward the chemical and physical processes responsible for removing nitrogen oxides from the atmosphere.

Determination of the extent to which nitrogen oxides are removed from an air mass and the chemical and physical processes by which this removal occurs is complicated by continuous and variable NO_x imputs and variable dilution of the air mass. These problems were largely circumvented in this study through the use of a gaseous tracer, namely carbon monoxide*. The use of a rather inert tracer such as CO allows one to compute the fraction of nitrogen oxides removed from the air mass. It is then possible to attempt a "balance" between nitrogen oxide reactant and nitrogen-containing reaction products. The quality of the nitrogen balance and the nature of the reaction products are then quite suggestive as to the extent of NO_x removal from the atmosphere and the chemical and physical mechanisms of removal.

In order to derive a "balance" between nitrogen oxide reactant and NO_x reaction products, it is necessary to determine the fraction of NO_x which has been converted to products or removed from the atmosphere at any given time. This "NO_x Loss" is derived using the equation

$$NO_x \text{ Loss} = \frac{(CO)_{measured}}{(CO/NO_x)_{emission\ inventory}} - (NO_x)_{measured} \qquad (1)$$

where $(CO/NO_x)_{emission\ inventory}$ is the ratio based on emissions inventory estimates and $(CO)_{measured}$ and $(NO_x)_{measured}$ are the actual concentrations of CO and NO_x at any point in time. The first term on the right-hand side of the equation can be thought of as a predicted NO_x concentration based on the estimated emissions ratio of CO and NO_x and the measured CO concentration at any given time. The assumption is made here that over short time periods (several hours) CO is inert and can be used as a tracer for NO_x concentration. Subtracting the actual measured NO_x concentration from the predicted concentration yields the term "NO_x Loss". This calculated NO_x-loss term can be compared to the concentration of nitrogen-containing reaction products such as PAN and nitric acid in order to determine a nitrogen balance.

---

*CO data taken simultaneously with our measurements were kindly supplied by EPA and LAAPCD.

The use of Equation 1 as it stands would lead to an over-estimate of NO_x loss due to the presence of background CO, which is in no way associated with urban emissions sources. For the most accurate determination of "NO_x loss" we must eliminate any CO contribution which is not associated with the emissions inventory estimates. A convenient way of doing this involves rearrangement of Equation (1) as follows:

$$\frac{(CO/NO_x)_m}{(CO/NO_x)_{E.I.}} - 1 \quad (NO_x)_m = NO_x \text{ loss} . \tag{2}$$

The equation in this form makes use of the measured CO/NO_x ratio, which we obtain from the slope of the CO versus NO_x regression line. The slope of this regression line eliminates the CO intercept, which contains CO contributions that are invariant to changing source emissions--e.g., background CO. The intercept can be viewed as CO from a source not associated with NO_x. Since all the important CO sources (with auto exhaust by far the most important in urban areas) are associated also with NO_x emissions, extrapolating NO_x to zero (the CO intercept) yields an average CO concentration which is not included in the emission inventory and which, therefore, must be excluded from the "NO_x loss" computation. The slopes of the CO vs NO_x regression lines for St. Louis and West Covina are shown as $(CO/NO_x)_m$ in Table VI. Also shown in the table are the emissions inventory ratios, $(CO/NO_x)_{E.I.}$, and the average calculated NO_x losses.

Table VI. Average NO_x Loss in St. Louis and West Covina

	$(CO/NO_x)_m$	$(CO/NO_x)_{E.I.}$	NO_x Loss, ppm
St. Louis	14.6 ± 3.2	14.6	0.000 ± 0.012
West Covina	18.4 ± 0.6	14.3	0.035 ± 0.006

It is apparent from the data in Table VI that the average NO_x loss in St. Louis, 0.000 ± 0.012 ppm, is small in comparison with the average concentration of NO_x. The average sum of PAN and HNO_3 over the same time period was 0.007 ppm, well within the 0.012 ppm deviation. Thus, the average loss of NO_x in St. Louis is quite small. The use of acetylene as a tracer has also confirmed this finding[2].

The average calculated NO_x loss in West Covina as reported in Table VI, is 0.035 ± 0.006 ppm. The 0.006 ppm deviation is merely the statistical deviation about the slope calculation. There are several other sources of error however, which may have a much greater impact on the accuracy of the "NO_x loss" calculation for West Covina. First, the calculation depends on emission inventories which were over two years old at the time of this study. Second, the CO data from West Covina were obtained by NDIR, a procedure which is less sensitive and more susceptible to interference than the gas chromatographic procedures used in St. Louis. The effect of these interferences is presumably largely eliminated in our calculation by the use of the slope of the CO vs NO_x regression line. The intercept of the regression line now contains a background CO component and a component caused by instrumental insensitivity and interference. However, the computed NO_x loss is still likely to be overestimated because of these sources of error. Thus we cannot at this time make an unqualified judgment as to the balance between NO_x reactants and products in West Covina due to the many factors which may be influencing the accuracy of the "NO_x loss" calculation. We can say however, that the "NO_x loss" appears to be a small fraction of the total NO_x concentration in West Covina. A much more exact determination of NO_x loss will be carried out in the near future using CO data collected by gas chromatography in West Covina simultaneously with our study.

One final view of the problem can be gained by examining the time dependence of the composited NO_x loss profile. This 5-week averaged plot is shown in Figure 3. The relative concentration shown on the ordinates of these plots may be thought of as parts per hundred million (pphm). It should be observed first of all that we have allowed the data to form their own baseline. Our detailed calculation indicates that this baseline may be obscuring an apparent 20 ppb NO_x loss which remains constant with time (shows no diurnal variation). This apparent 20 ppb loss may be due to NO_x removal by dry deposition processes, it may represent inaccuracies in the emissions inventories, or it may represent some other source of undefined error. The fact that it is constant however, indicates that, if it is indeed a real loss, it is probably not photochemical in nature.

Of the two prominent humps displayed in the "NO_x loss" curve, the second one can be accounted for entirely by the sum of PAN and HNO_3. This is illustrated by the profile in the lower portion of the figure. We can say, therefore, that the mechanism of afternoon loss of NO_x is photochemically related and that the magnitude of the loss can be completely accounted for by measured NO_x reaction products.

The explanation for the apparent early morning loss of NO_x is unclear at this time. Obviously, for the "NO_x loss" to increase as shown in the figure, the $(CO/NO_x)m$ ratio in Equation (2) must increase. Since this early morning increase in the "NO_x loss" curve appears at the same time that the CO and NO_x emissions sources are undergoing their most dramatic change of the day, the possibility exists that the early morning "loss"

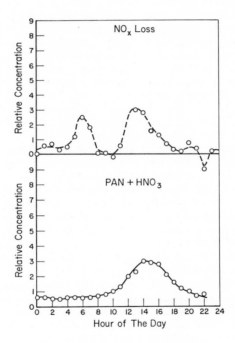

Figure 3. Composite profiles, West Covina

is artificial. The rationale is as follows: A major increase
in the auto exhaust contribution to the air mass occurs between
5:00-8:00 a.m., judging from the average NO_x profile in Figure
2. Since auto exhaust has a considerably higher CO/NO_x ratio
than the normal Los Angeles basin mixture (approximately 24
versus 14.3), the morning "NO_x loss" peak may only be the result
of a different emissions mix during peak traffic hours and not
truly reflect removal of NO_x from the air mass. Indeed, if the
additional morning auto exhaust burden (at a CO/NO_x ratio of 24)
raises the normal CO/NO_x emissions ratio from 14.3 to 16.5, then
the 6:00 a.m. apparent "NO_x loss" peak would be completely
eliminated. At this time there is no sure way to incorporate
a variable emission inventory ratio into our calculations,
although it seems very likely that the emissions ratio must
vary during the day due to traffic patterns and variations in
other emissions sources. At this point we can only suggest the
variation in the CO/NO_x emissions ratio as a probable cause of
the apparent morning NO_x loss.

Conclusions

    Several tentative conclusions can be drawn from our results.
Since detailed discussions of the data supporting each of these
conclusions has been presented elsewhere[2], we will simply
state our major observations and conclusions here:

   (1) Based on our "NO_x Loss" calculations it appears
        that the mechanisms by which the nitrogen oxides
        are removed from the atmosphere involve rela-
        tively slow processes.

   (2) In West Covina, the greatest loss of nitrogen
        oxides was observed during midafternoon on
        high photochemical smog days. This portion
        of the "NO_x Loss" profile was highly corre-
        lated with ozone concentration. If certain
        assumptions are accepted, the afternoon or
        photochemical loss of nitrogen oxides appears
        to be largely accounted for by the sum of the
        PAN and nitric acid concentration.

   (3) There was some evidence that nitric acid is
        removed from the air by alkaline-surface
        glass-fiber filters, but not by high-purity
        quartz-fiber filters. If true, this would
        mean that much particulate nitrate data
        collected over the years may have been
        strongly influenced by gaseous nitric acid.

   (4) There was some indication from our data
        that ozone is advected into the St. Louis
        region during early morning hours. While
        the source of this ozone is not yet clear,
        there is some evidence that PAN and possibly

nitric acid are associated with the nighttime ozone.

The program is continuing and the second year effort will
be devoted in part to further validation and documentation of
our nitric acid measurements and in part to further analysis
and interpretation of our first-year field results.

Acknowledgments

The work reported here was supported by the Coordinating
Research Council, Inc., and the U.S. Environmental Protection
Agency. We would like to acknowledge the valuable contributions
to this study made by the members of the CRC-APRAC committee,
CAPA-9-71; E. S. Jacobs, J. J. Bufalini, E. H. Burk, W. A.
Glasson, R. Hammerle, and the late D. Hutchinson.

Literature Cited

1. Spicer, C. W. and Miller, D. F., "Nitrogen Balance in Smog
   Chamber Studies", to be published in J. Air Poll. Control
   Assoc.; presented at 67th Annual Meeting, Air Poll. Control
   Assoc., Denver, 1974.

2. Spicer, C. W., "The Fate of Nitrogen Oxides in the Atmosphere", Battelle-Columbus report to EPA and CRC,
   September, 1974.

3. Miller, D. F. and Spicer, C. W., "A Continuous Analyzer
   for Detecting Nitric Acid", to be published in J. Air
   Poll. Control Assoc.; presented at the 67th Annual Meeting,
   Air Poll. Control Assoc., Denver, 1974.

15

High Ozone Concentrations in Nonurban Atmospheres

LYMAN A. RIPPERTON, JOSEPH E. SICKLES, W. CARY EATON, CLIFFORD E. DECKER, and WALTER D. BACH

Research Triangle Institute, Research Triangle Park, N. C. 27711

The National Ambient Air Quality Standard for photochemical oxidant[1] was set in 1971 as < 0.08 ppm hourly average, not to be exceeded more than once a year. Until the mid-1960's, non-urban ozone (O_3) concentrations near the surface of the earth were considered to be in the 0.00 to 0.06 parts per million (ppm) range, with the average close to 0.02 ppm.[2] Since the mid-1960's, a rapidly growing list of reports cites O_3 concentrations of 0.08 ppm or above in small towns and in rural locations.[3]

Considerable difference of opinion exists regarding the origin of these high concentrations of O_3 in rural and semi-rural areas. Basically, the explanations advanced for the phenomenon are:

1. The O_3 has descended from the stratosphere to the earth's surface.

2. The O_3 has been synthesized in the troposphere from only natural precursors in photochemical primary and secondary reactions.

3. The O_3 has been synthesized in urban environs and has been transported to rural areas without further significant chemical reaction.

4. The O_3 has been synthesized in transit to and beyond the rural sampling site in systems originating in urban areas (with or without additional injection of precursor material into the moving system).

In the summers of 1970,[4] 1972[5] and 1973,[3] the Research Triangle Institute (RTI) made field studies under Environmental Protection Agency (EPA) sponsorship, which have helped characterize the phenomenon of high nonurban concentrations of O_3. The following report is largely taken from the 1973 study.

Studies of experimental air pollution are being carried out in transparent bags and in smog chambers to aid in the interpretation of field observations.

Procedure

Field Studies. In the summer of 1973, RTI set up field stations at or near Lewisburg, West Virginia, McHenry, Maryland, Kane, Pennsylvania and Coshocton, Ohio to take measurements of ambient concentrations of O_3, and oxides of nitrogen (NO–NO$_2$). Nonmethane hydrocarbon data proved invalid. Only O_3 was measured at McHenry. Figure 1 shows the location of the four sampling stations. A few measurements of carbon monoxide (CO) were made and some grab samples were taken for subsequent gas chromatographic analysis of hydrocarbons. Chemiluminescent techniques were used for O_3 and NO_x measurements. Carbon monoxide and individual hydrocarbons were analyzed by gas chromatography performed by personnel of EPA.

An aircraft equipped with a chemiluminescent meter was used to make O_3 measurements aloft, between, and over the fixed field stations.

Mean wind data at 900 mb from NOAA were used for trajectory analysis.

Experimental Studies. Clear FEP Teflon bags with low concentrations of isopentane and NO$_2$ (see Table 1 for concentrations) were exposed to sunlight starting about 0830 EDT. Ozone concentrations were measured intermittently until sundown.

Isopentane ppmC / NO$_2$ ppm	1.00	0.50	0.25
0.25	0.13	0.11	0.10
.010	0.08	0.06	0.06
.005	0.06	0.02	0.03

Table 1. Bag Study, Ozone Synthesis, Isopentane + NO$_2$ in Sunlight (~0830 – 1500 EDT, March)

Experimental studies are being carried on in smog chambers which are designed to help explain the high rural O_3 phenomenon. Results of one such test, which was performed after the original presentation of this report, are presented briefly as they have an important bearing on the conclusions drawn from field data.

The chamber is 988 cubic feet in volume, made of 5 mil FEP Teflon, which covers an interior aluminum frame. A 60 hour static run was made with an initial charge of 3.0 ppmC hydrocarbon of a mixture of alkanes and alkenes, 0.1 ppm NO_x, 40% NO_2, exposed to sunlight for 2 1/2 diurnal cycles starting at sunup. Chemiluminescent measurements were made of O_3 and NO_x. Nitrogen dioxide was also measured, intermittently, with a Saltzman technique.

Results

The hourly averages of O_3 for each of the four fixed stations are graphed in Figure 2. In RTI's 1973 summer study, the NAAQS was exceeded during:

15% of 1663 hours at Lewisburg
37% of 1652 hours at McHenry
30% of 2131 hours at Kane
20% of 1785 hours at Coshocton

Concentrations of NO_2 are shown in Figure 3. Selected aircraft data from vertical flights are shown in Figures 4 and 5.

Figure 6 is an abstract of a trajectory study. Figure 6 shows the position 24 hours earlier of those trajectories reaching Kane, Pennsylvania which had an average 12-hour O_3 concentration equal to or greater than 0.08 ppm. Most of these originated to the west, but all trajectories, including those of air with little O_3, tended to originate to the west of Kane. Figure 6 shows a gradient of 1° lines of latitude and longitude in the areas studied in 1973.

Table 1 reports results of the isopentane NO_2 irradiations. The results of the smog chamber run are abstracted in Figure 7.

Discussion

In the following paragraphs the authors have discussed the implications of the observational data and attempted a crude model, or paradigm describing the system.

Atmospheric layering phenomena involving O_3 concentration near the ground are complex, but the data in Figures 4 and 5 show considerably higher concentrations at ground level at Lewisburg than simultaneous concentrations at aircraft level. At Kane, at ground level at 1130 EDT on August 8, the O_3 concentration was 0.13 ppm while at 2000 feet above the surface, the simultaneous concentration was about 0.08 ppm. This is evidence that the ground level O_3 had not subsided from the stratosphere and was generated near the ground.

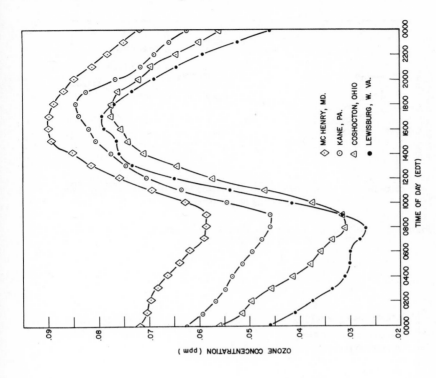

Figure 2. Mean diurnal ozone concentrations, June 26–September 30, 1973

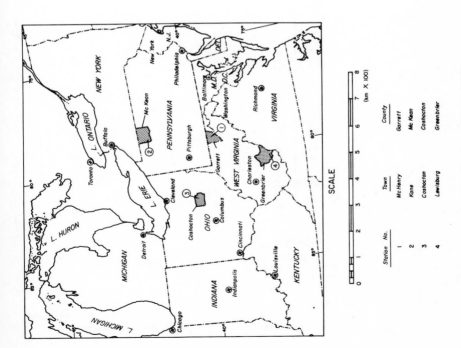

Figure 1. Ozone monitoring stations, 1973

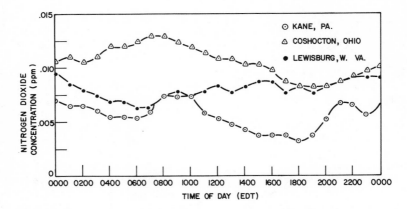

Figure 3. Mean diurnal nitrogen dioxide concentrations, June 26–September 30, 1973

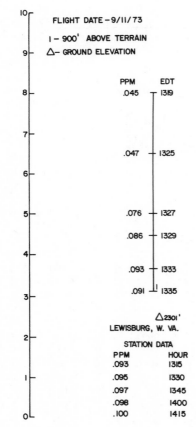

Figure 4. Vertical ozone measurements, September 11, 1973

It should be noted also that the large percent of time that
the high concentrations existed and the period of time (nearly
four months) that high concentrations were observed argues
against a stratospheric origin of the ozone reported in this
study.

Although only nine hydrocarbon samples were taken at the
stations in the course of the 1973 study, acetylene was found in
all of them, at concentrations ranging from 2 to 63 parts per
billion (ppb). It is well known that acetylene has no known
sources other than anthropogenic.(6)   The CO concentrations in
the samples ranged from 0.27 to 0.46 ppm. Although natural background varies, it is generally considered to range between .08
to .12 ppm. These data indicate the presence of anthropogenic
pollution. The measured NO_2 concentrations were low, 0.005 to
0.030 ppm, but are still greater than those considered as background, about 0.008 ppm. This is another indication of anthropogenic sources. These data reveal the constant presence of some
anthropogenic pollution which refutes the argument that the synthesis of high concentrations of O_3 is accomplished from natural
precursors only.

It should be pointed out that natural emissions of both organic vapors and of oxides of nitrogen(7) occur and these will
enter into the ozone generative processes and participate in the
same kinds of photochemical reactions as their anthropogenic
counterparts.(8)

The transport of O_3, especially aloft, within a few thousand
feet of, but out of touch with, the ground, has been demonstrated.
(9)   However, the fact that a location like Indio (population
about 14,000) can exceed the NAAQS more frequently than any other
sampling site in California(10) (e.g., Los Angeles, Riverside)
is clear proof that O_3 synthesis does not stop at the city limits
but continues as the photochemical system travels downwind.

If it is admitted that the O_3 measured in high concentrations in our field synthesized in the troposphere, the large
question that remains is: if the precursors are not all natural,
where do they come from and does anthropogenic pollution have a
special role in generating the high nonurban concentrations of
O_3?

Although Figure 6 shows that the trajectories of air ar-
riving at Kane, Pennsylvania in the summer of 1973 which could
bring high concentrations of O_3, more trajectories from the same
locations noted on the map had low concentrations of O_3 (i.e.,
< 0.08 ppm) than high. Figure 6 also shows that the trajectories
of high O_3 air came from almost all possible origins. (Few trajectories of any concentration originated to the east of Kane.)
This is indicative of an area-wide system of pollution sources
rather than the plume of a specific city.

The data from the bag study presented in Table 1 show that
O_3 can be synthesized in photochemical systems with low concentrations of NO_x (e.g., 0.01 ppm) and with hydrocarbon vapors with

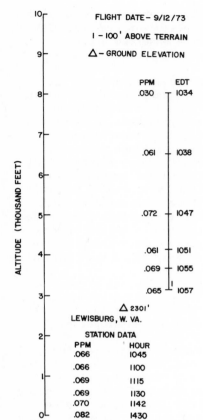

Figure 5. Vertical ozone measurements, September 12, 1973

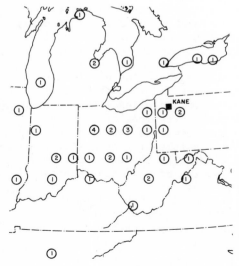

Figure 6.   Number of occurrences, by position 24 hr earlier, of air with ozone concentrations > 0.08 ppm (12 hr average), Kane, Penn., summer 1973

little or no O_3-reactive character. A high hydrocarbon to NO_2 ratio also seems to be necessary. A concentration of 0.01 ppm NO_2 produced 0.08 ppm of O_3 only when the isopentane-carbon to NO_2 ratio reached a value of 100:1.

One importance of the smog chamber profile seen in Figure 7 is that it shows that the O_3 concentration of the chamber did not drop to zero at night. Furthermore, on two consecutive days the chamber mix generated an increment of O_3 greater than the 0.08 ppm of the NAAQS. This generation took place in the presence of measured concentrations of NO_x lower than most reported from field studies of the high rural O_3 problem.

Integrating the information found in the literature with concentration numbers and ratios generated by the 1973 RTI field study, the authors have formed a tentative view of the processes and sequence of events which lead to higher than expected O_3 concentration in rural areas. Although these arguments are advanced for the occurrence of high O_3 concentrations in the north-central and north-eastern U.S., many of these arguments could apply to other geographical areas.

Proposed System Behavior

High pressure systems often sweep from the northern plains, south of the Great Lakes, across the east to the North Atlantic seaboard and out to sea. Meteorological factors in atmospheric high pressure systems are best for containment of pollutants and buildup of their concentrations. It is postulated that over the plains, with a low population density and emission rate, relatively low precursor concentrations and oxidant concentrations exist. As the high pressure system moves it picks up natural precursors as well as anthropogenically generated precursors. Precursor input coupled with the relatively low interchange rates of precursor material to the ground lead to a quasi-steady state system of O_3 generation in the air mass. This urban-natural mix would have a greater concentration of stable intermediates (e.g., HONO, CH_2O), and in sunlight a greater capacity to produce •OH radicals than natural air alone.

When the air-mass moves over a region of high pollution, i.e., an urban-industrial area, anthropogenic precursors are injected. Initially, these are in relatively high concentration in the vicinity of the injection region, but they dilute downwind. This has two important consequences. First, the urban-industrial plume is only a perturbation of the relatively wide-area pollution system. This means that specific source identity is lost at a sufficient distance downwind. Second, the initially high concentrations of pollutants are diluted with distance.

It is a well-known phenomenon that reducing the concentration of precursors, either initially(11) or in the course of a smog chamber run(12) can lead, under some conditions, to an enhanced net O_3 generation. In computer simulations with an O_3

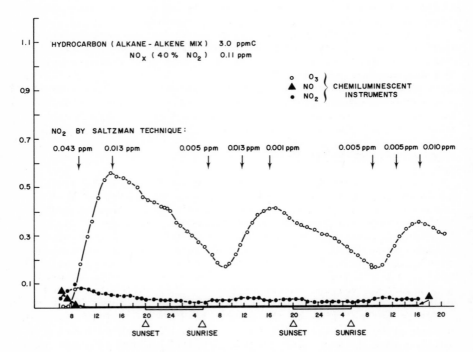

Figure 7.   Sixty-hour outdoor smog chamber run, June 3, 4, 5, 1975

generation model derived by Sickles and Jeffries, the generation
of both stable intermediates (e.g., RCHO and HONO) which are
capable of providing •OH radicals and various radicals themselves
was also enhanced by dilution. This has relevance to a system
of air pollution moving out of a city and being diluted with
nonurban air.

Ozone is presumed to be generated by the following types of
reactions.(13) (See Table 2.)

In the case in which aromatic and alkane hydrocarbons predominate over alkenes (as in a "spent" photochemical system) a
source of •OH radicals at the beginning of daylight would speed
up reactions so that a greater realization of the oxidant potential of a hydrocarbon-NO_x system could be realized in a shorter
time. This could be important in keeping O_3 concentrations high
in urban-natural mixtures of air.

Note in Figure 7 that the daily maximum hourly average O_3
concentration moved closer to sunset after the first day of irradiation. This provides an analogy with a "spent photochemical
system" [or nearly spent] in the free atmosphere. (Los Angeles
oxidant generally peaks shortly after noon. Peak ozone concentrations in the RTI field study of 1973 tended to be at 1700 or
later.)

In the field studies the diurnal trend of O_3 concentrations
occurred most of the time, but high nighttime concentrations
occurred from time to time. The authors wish to postulate that
high night time concentrations represent a reservoir of high concentrations existing aloft within a few thousand feet of the
ground. This would represent O_3 left over from the previous day
in a "spent" photochemical system in which O_3 destructive agents
were of low concentration (or of low reactivity). A nocturnal
decrease in O_3 concentration probably accompanies a radiation
inversion with surface O_3 destroyed by the ground and by local
pollution emissions. In the morning surface concentrations of
O_3 increase due to both O_3's transfer from the reservoir aloft
and to synthesis involving the local emissions and admixed stable
precursors from aloft.

It is postulated that in a rural area the local emissions
are usually not great enough alone to account for the increase
seen in the upswing of the diurnal cycle. It is also postulated
that natural emissions are generally not great enough alone to
generate the concentrations of O_3 observed.

The results of the isopentane-NO_2 study and the smog chamber
run are strong indicators that when an O_3 generative system does
not have much O_3 destructive ability little NO_x is needed to produce large concentrations of O_3. A question which this raises
is, "What keeps O_3 from being generated in concentrations > 0.08
ppm all the time when the sun is shining?"

A partial answer will probably be found when natural systems
can be properly described and their chemistry characterized as to
O_3 generative and destructive reactions. Since probably the same

Ozone Generation and Ozone-NO$_x$ Relationships

$$NO_2 + h\nu \longrightarrow NO + O$$

$$O + O_2 + M \longrightarrow O_3 + M$$

$$NO + O_3 \longrightarrow NO_2 + O_2$$

No Net O_3

NO Oxidation (Non-O$_3$)

$$NO + ROO \longrightarrow NO_2 + RO\cdot$$

$$\overset{O}{RCOO\cdot} + NO \longrightarrow NO_2 + \overset{O}{RCO\cdot}$$

$$HO_2\cdot + NO \longrightarrow NO_2 + HO\cdot$$

One O_3 Netted for Each NO Oxidized

Peroxy-radical Formation

$$RH + \cdot OH \longrightarrow R\cdot + H_2O$$

$$\overset{O}{RCH} + \cdot OH \longrightarrow \overset{O}{RC\cdot} + H_2O$$

$$R\cdot + O_2 \longrightarrow ROO\cdot$$

$$\overset{O}{RC\cdot} + O_2 \longrightarrow \overset{O}{RCOO\cdot}$$

Selected Hydroxyl Radical Formation Schemes

1. $NO + NO_2 \xrightarrow{\;H_2O\;} 2HONO$

    $HONO + h\nu \longrightarrow NO_2 + \cdot OH$

2. $H_2C=O + h\nu \longrightarrow H\cdot + \overset{O}{HC\cdot}$

    $H\cdot + O_2 \longrightarrow HOO\cdot$

    $\overset{O}{HC\cdot} + O_2 \longrightarrow HOO\cdot + CO$

    $HOO\cdot + NO \longrightarrow NO + \cdot OH$

3. $O_3 + h\nu \longrightarrow O(^1D) + O_2$

    $H_2O + O(^1D) \longrightarrow 2\cdot OH$

Table 2. Selected Reactions Involved in Ozone Generation

kinds of reactions are involved in generating the O_3 in the non-urban areas that are also operative in the urban atmosphere, this will be more a matter of relative rates than of new reactions. Field and chamber measurements of stable intermediates in the $HC-NO_x-O_3-h\nu$ system are eminently desirable for proper interpretation of such chemical relationships.

Conclusions

Air in the area of study (Ohio, Pennsylvania, Maryland and West Virginia) in which high O_3 concentrations were generated had evidence of anthropogenic contamination (other than O_3). This air appeared not to come always from a specific geographical location. The O_3 generative capacity of the system is judged by the authors to be an air mass characteristic.

The O_3 is postulated as being generated in a relatively "spent photochemical system." The system is referred to as "relatively spent" because there is evidence of continuous injection of a small amount of pollution. The authors have presented a rough paradigm in which the capacity of the so-called spent photochemical system to generate O_3 from low concentrations of NO_x (i.e., 5-20 ppb NO_x) is due to a reduced capacity of the system to destroy O_3 and an enhancement of the rate of O_3 accumulation due to reactions involving stable intermediates which provide a source of $\cdot OH$ radicals. The role of contaminated city air is to provide a source of the stable intermediates greater than would occur in nature.

Literature Cited

1. Federal Register (April 30, 1971), 36 (No. 84), 8187.
2. Junge, C. E., "Air Chemistry and Radioactivity," pp. 37-59, Academic Press, New York (1963).
3. Research Triangle Institute, "Investigation of Ozone and Ozone Precursor Concentrations at Nonurban Locations in the Eastern United States," Research Triangle Park, N.C. (1974). (Also issued as Environmental Protection Agency Report No. 450/3-74-034.)
4. Richter, H. G., "Special Ozone and Oxidant Measurements in the Vicinity of Mount Storm, West Virginia," RTI Task No. 3, NAPCA Contract No. 70-147, Research Triangle Park, N.C. (1970).
5. Research Triangle Institute, "Investigation of High Ozone Concentrations in the Vicinity of Garrett County, Maryland and Preston County, West Virginia," Research Triangle Park, N.C. (1973). (Also issued as Environmental Protection Agency Report No. 450/3-75-036.)
6. Miller, S. A., "Acetylene: Its Properties, Manufacture and Uses," Academic Press, New York (1965).

7. Robinson, E. and R. C. Robbins, "Sources, Abundance and Fate
    of Gaseous Atmospheric Pollutants," Stanford Research Insti-
    tute Project PR-6755, prepared for American Petroleum Insti-
    tute, New York (1968).
8. Ripperton, L. A. and D. Lillian, "The Effect of Water Vapor
    on Ozone Synthesis in the Photo-oxidation of Alpha-Pinene,"
    J. Air Poll. Control Assoc. (1971) 21:629-635.
9. Lea, D. A., "Vertical Ozone Distribution in the Lower Tropo-
    sphere Near an Urban Pollution Center," J. Appl. Meteor.
    (1968) 7:252-267.
10. Maga, John, California Air Resources Board, personal commu-
    nication.
11. Niki, H., C. E. Daby, B. Weinstock, "Mechanics of Smog For-
    mation," preprint of Fuel Science Department, Ford Motor
    Corporation, presented 101st National Meeting, American
    Chemical Society, Los Angeles, California (1971).
12. Fox, D. L., R. Kamens, H. E. Jeffries, "Photochemical Smog
    Systems: Effects of Dilution on Ozone Formation," Science
    (1975) 188:1113-1114.
13. Demerjian, K. L., J. A. Kerr, J. G. Calvert, "Mechanism of
    Photochemical Smog Formation," in Advances in Environmental
    Science and Technology, Vol. 4, pp. 1-262, J. N. Pitts,
    R. Metcalf, eds, John Wiley, New York (1974).

Gas-Phase Reactions of Ozone and Olefin in the Presence of Sulfur Dioxide

D. N. MCNELIS, L. RIPPERTON,[a] W. E. WILSON,[b] P. L. HANST,[b] and
B. W. GAY, JR.[b]

Environmental Protection Agency, Las Vegas, Nev. 89114

The loss of visibility in atmospheres laden with photochemical smog is largely attributed to the presence of sulfate
aerosols formed in the oxidative consumption of sulfur dioxide.
This facet was thought to be the most objectionable effect of the
sulfate species and is reflected in the establishment of the
National Ambient Air Quality Standard for sulfur dioxide. A recent review, however, has indicated that the adverse health
effects appear more strongly associated with suspended particulate sulfate than with the sulfur dioxide (1).

The reaction between ozone and sulfur dioxide is too slow to
be significant at atmospheric levels, but the reaction of ozone
with certain unsaturated hydrocarbons is considerably faster.
The products of this reaction include an unspecified but highly
reactive species which has been advanced as a potential oxidant
for the sulfur dioxide (2), (3). A reaction of this type could
conceivably account for the aerosol formation observed in the
photochemical system.

This process, along with those of direct photochemical oxidation and catalytic oxidation of sulfur dioxide, is for the most
part poorly understood and becomes considerably more complex
when in the presence of other air pollutants. In recent years,
major advances in the knowledge of homogeneous gas phase reactions have been accomplished although the factors governing the
gas to particle conversion remain largely unresolved and are only
briefly discussed in published reviews of the chemistry of air
pollution.

Because investigators are unable to control or isolate
the various gaseous components or environmental factors leading
to aerosol production in the ambient environment, many of the
investigations designed to study both the gas phase mechanisms
and the aerosol formation are conducted in chambers under controlled conditions. It was in such a system that the dark phase
reaction of olefin-ozone-sulfur dioxide was investigated in an

a. Present Address: Research Triangle Institute
    Research Triangle Park, North Carolina 27711
b. Present Address: Environmental Protection Agency
    Research Triangle Park, North Carolina 27711

attempt to elucidate the mechanism involved in the oxidative
consumption of sulfur dioxide with concomitant aerosol formation.
The effect of several variables on the reaction stoichiometry
and on the aerosol production was also investigated. These
variables included the reactant concentrations, the relative
humidity, molecular oxygen concentration and the olefin species,
although propylene was the primary olefin studied. A discussion
of the previous work related to this study as well as the results
of the aerosol aspects is included in an EPA report (4).

Experimental Arrangement and Procedures

A photograph of the major portion of the instrumentation
used routinely during this study is shown as Figure 1. The reactor is shown at the top center and the aerosol instrumentation
directly below, except for the integrating nephelometer. The
reactant introduction system can be seen in the background near
the center of the picture. The gas chromatograph is shown on
the right and the sulfur dioxide and ozone analyzers are to the
left of the aerosol instrumentation table. At the extreme left
is a computer and display scope used for the processing of data
from the optical particle analyzer. The reaction bag was
constructed of Tedlar PVF film and when fully inflated contained
approximately 437 liters. An aluminized polyester bag was constructed and placed over the Tedlar to eliminate the transmission
of UV radiation and also to reduce the rate of permeation of
gases and vapors into or out of the reactor.

Figure 2 shows the Fourier Transform Infrared Spectrometer.
The folded path cell is in the background with the detectors,
interferometer and recorder in the foreground.

The mean temperature for the experimental runs was 25°C and
the two levels of relative humidity used were 22 and 38%. The
initial concentration of propylene ranged from about 0.9 to 8.6
ppm and for ozone the range was 0.3 to 2.8 ppm. The initial sulfur dioxide concentration was generally, at one of three levels,
0, 0.2 or 0.6 ppm.

Results and Discussion

The variation of concentration with time for a typical propylene-ozone-sulfur dioxide run is shown in Figure 3. The near
1:1 stoichiometric consumption for the two principal reactants is
readily apparent in the nearly constant difference in their concentrations with time. The consumption of the sulfur dioxide is
markedly slower than it is for the other two reactants. The
acetaldehyde production is also shown and, as it does not appear to follow linearly the loss of the propylene, the process
leading to its appearance is more complex than if it were
formed in a single bimolecular reaction.

The data obtained from the series of experiments were used

Figure 1.   *Laboratory arrangement for olefin-ozone-sulfur dioxide studies*

Figure 2.   *Fourier transform infrared spectrometer*

to plot the variation in the apparent stoichiometry of the olefin-ozone reaction with the initial reactant concentration ratio (Figure 4).

Thirteen of the 22 data points plotted represent ozone-propylene reactions conducted in air. Ten of these 13 experiments were conducted in the presence of sulfur dioxide. The presence of the sulfur dioxide had no apparent effect on the stoichiometry of the primary reaction as all of the data closely follow a smooth curve. This result indicates that the oxidative consumption of the sulfur dioxide is due to a reactive product of the propylene-ozone reaction rather than to a primary reactant such as ozone. This conclusion was fully substantiated through an experiment in which ozone and sulfur dioxide were the only reactants. Consumption of those species was so slow that their loss could not be distinguished from losses due to wall effects.

The data also indicate that the system is complicated by secondary reactions. The apparent stoichiometry of the reaction is decidedly affected by the initial concentration ratio. Generally, in a system which initially has propylene in excess, the consumption ratio of the propylene to ozone is greater than one while in a system which initially has an ozone excess condition the value of this ratio is less than one.

The molecular oxygen concentration was sharply reduced for several runs in an attempt to determine whether oxygen has a role in the reaction. The results of four of these experiments in which nitrogen was used as the diluent gas are recorded as triangles on Figure 4. An obvious reduction in the propylene to ozone consumption ratio occurred, indicating that molecular oxygen plays a significant role in modifying the consumption of one or both of the primary reactants.

Observation of this molecular oxygen effect led to the postulate that a secondary attack on the propylene is occurring in this reaction. As the consumption ratio changes in the presence or absence of molecular oxygen, a secondary attack on the olefin by the reactive intermediate can be eliminated as being responsible for the stoichiometric change. Other reactions which could account for the oxygen effect and the deviation from 1:1 stoichiometry are those which involve an interaction of the intermediate species and molecular oxygen. The products of three of these type of reactions which have been postulated include the hydroxyl, peroxyacetyl, methyl and the hydroperoxy radicals and ozone. Each of the reactions results in the production of a species capable of further interaction with the olefin. Some of these species also react with carbon monoxide although the rate constant for the reaction of the hydroxyl radical with carbon monoxide is approximately 10^6 times greater than the reaction with the hydroperoxy radical and about 10^{12} greater than the reaction of carbon monoxide with ozone. The addition of an appropriate concentration of carbon monoxide to serve as a scavenger in the system should provide a convenient method of testing for the

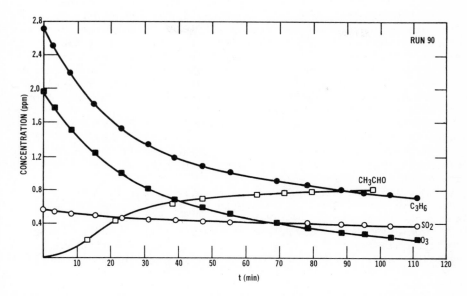

Figure 3.   Variation of concentration with time for propylene-ozone-sulfur dioxide reaction

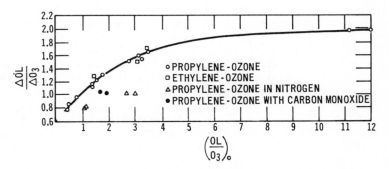

Figure 4.   Stoichiometry of olefin-ozone reaction vs initial reactant concentration ratio

formation and presence of the hydroxyl radical. As the hydroxyl radical would also react with the olefin and sulfur dioxide in the system in addition to any aldehydes formed and other minor products of the olefin reaction, a sufficient carbon monoxide concentration was needed to significantly affect the overall reaction stoichiometry. It was calculated that for an initial propylene concentration of 3 ppm, a carbon monoxide concentration of about 300 ppm would cause the loss of the hydroxyl radical in a reaction with the carbon monoxide to be equal to the loss from the total of the other reactions.

Two experiments in which 458 ppm of carbon monoxide was added to the system were conducted and the stoichiometric results substantiate the postulate relative to the formation of the hydroxyl radical in this reaction and its subsequent secondary attack on the olefin species.

One final result included graphically on Figure 4 is the lack of an effect due to a change in the relative humidity over the range studied.

The data show that the stoichiometry of the propylene-ozone reaction was generally other than 1:1. Because the reactant consumption ratio dropped when the partial pressure of the molecular oxygen was reduced, it was speculated that ozone and/or the hydroxyl radical were being produced in the reaction of ozone with propylene. The reduction of the reactant consumption ratio when the carbon monoxide was added enforced the speculation relative to the formation of the hydroxyl radical which was subsequently scavenged by the carbon monoxide. These measurements do not, however, demonstrate whether the change in stoichiometry with the changing propylene/ozone initial condition is due to a production of propylene, enhanced consumption of ozone or to an interaction of these species with reaction products.

To investigate further the reaction mechanism with regard to the stoichiometry, the variation of the function of the consumption rate of one of the principal reactants over their product with time was noted. The variation of these two functions with time for one run is shown on Figure 5. If the only reactions involving these species were with each other, then the rate of olefin disappearance would equal the rate of ozone disappearance and the plots of the rate over reactant product versus time would yield a straight line of zero slope. The value of this function would be equal to the second order rate constant. If, however, a plot of one of the functions were to initially display a negative slope, then that species is being produced in the reaction. Conversely, if a plot of one of the functions were to display a positive slope initially, then that species is being consumed in another reaction. The resultant curves are typical for the series of experiments conducted. The values of the functions at the reported zero time are approximately equal and represent the rate constant for the propylene-ozone reaction. The mean and one standard deviation for k_1 computed from the 13 runs conducted in air and based on the rate of change in the ozone concentration was 1.40 ± 0.38 $(10^{-2}$ ppm^{-1} min$^{-1})$.

The curve which involves R_{O_3} is observed in the figure to increase with time suggesting that ozone is being consumed by a product or products of the principal reaction. Based on this result, the postulated generation of ozone by a secondary reaction is either nonexistent or is completely masked by the enhanced ozone consumption.

The other curve indicates that propylene is also being subjected to a secondary reaction. The postulate concerning the formation of the hydroxyl radical in a reaction involving the intermediate species with molecular oxygen can be used to explain that curve. As the hydroxyl radical is formed and reacts with propylene, the function plotted rises to its maximum value. Formaldehyde and acetyaldehyde are produced in the propylene–ozone reaction and eventually compete for the hydroxyl radical causing the function to diminish to some equilibrium value.

An inventory of the products formed in the gas phase reactions of propylene with ozone and propylene–ozone–sulfur dioxide was conducted by studying the infrared absorption spectra of the gaseous components in the reactor at a time when approximately 90% of the ozone had been reacted.

Ratio recordings measured by the Fourier Transform Infrared Spectrometer (FTS) for three spectral regions are shown in Figure 6 for runs 21 and 31. The infrared absorption path length for both of these measurements was 160 m through air at one atmosphere. The identified product compounds noted from the upper spectrum include carbon monoxide, formaldehyde, acetyaldehyde and a trace amount of formic acid. The lower spectrum is for an experiment conducted without any sulfur dioxide. Ketene was an additional product observed in that run.

Ratio plots with scale expansion of the central region for those same two runs are shown in Figure 7. The lower plot shows the ketene spectrum superimposed on the carbon monoxide continuum. The ketene is identified by the double indentation on the continuum with the right lobe characteristically being slightly broader and deeper. The dramatic disappearance of the ketene spectrum in the presence of sulfur dioxide shown on the upper plot indicates that the reaction rate of the oxidizing species with the sulfur dioxide is much faster than is the rate of the decomposition reaction. The formation of ketene in runs in an atmosphere with a sharply depleted molecular oxygen content was also noted.

An evaluation of data from the product analyses by the FTS system showed that the production of carbon monoxide was unaffected by the addition of sulfur dioxide to the system. Analogous to the postulated formation of ketene from the decomposition of the acetyl form of the intermediate species, carbon monoxide and water were speculated as forming from the decomposition of the formyl form of that species. For this analogy to hold, a higher concentration of carbon monoxide would be expected in the absence of

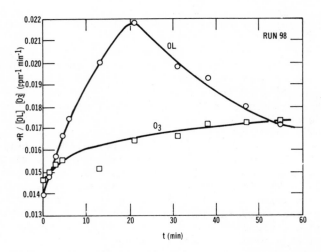

Figure 5.   Variation of the apparent rate constant with time due to secondary reactions

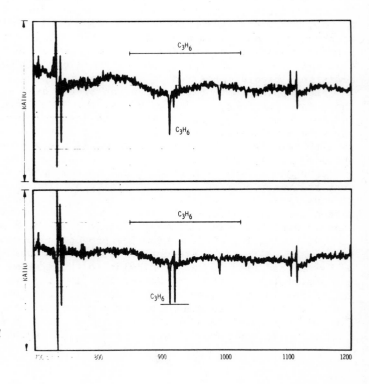

Figure 6.   Infrared
absorption spectra
of reactor contents

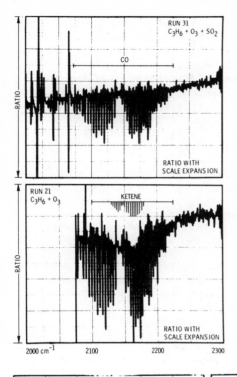

Figure 7. (left) Infrared spectra in region of carbon monoxide absorption

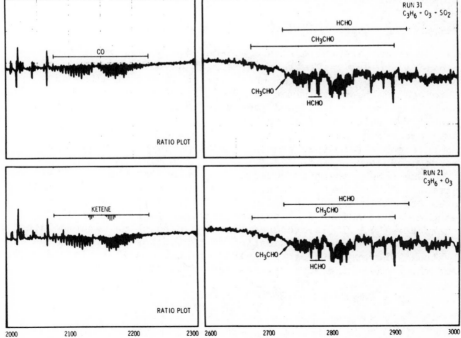

sulfur dioxide. This was not observed in these experiments as the production of carbon monoxide appeared constant relative to the rate of ozone consumption. Both forms of the intermediate species are operative in oxidizing the sulfur dioxide as evidenced by the formation of light scattering aerosols in both the ethylene and propylene systems. It had been postulated that carbon monoxide and methyl alcohol could be formed by still another decomposition reaction. At these initial reactant concentrations, no methyl or other alcohol was observed in these analyses and, therefore, such a reaction is either not operational or has a minor role in generating the carbon monoxide.

An increase in both formaldehyde and acetyaldehyde production was observed for the propylene-ozone thermal reaction when conducted in the presence of sulfur dioxide.

A product analysis of two propylene-ozone experiments conducted in a reduced oxygen atmosphere indicates that no formic acid was formed. This finding refutes the mechanistic step postulated in the literature in which the acid results from the decomposition of the formyl form of the intermediate species.

One further observation noted from these two experiments which made use of nitrogen as the diluent gas is the marked reduction in the production of both aldehyde species compared to the runs conducted in air. This result can be accounted for by another reaction of the intermediate species with oxygen resulting in the formation of ozone and aldehyde. It is significant to note that this reaction would contribute a major fraction of the aldehyde formed.

Tentative Reaction Mechanism

To summarize the evidence presented, the following scheme provides the salient features of a mechanism for the propylene-ozone-sulfur dioxide system which is consistent with the data observed in these experiments. The initial steps of the tentative mechanism are those which directly influence the concentration of either the sulfur dioxide or the species responsible for its oxidative consumption. The final steps in the mechanism are presented to account for other products observed in the various experiments conducted.

One scheme which is generally accepted for the ozonalysis reaction is the Criegee mechanism in which the reactive intermediates are the zwitterions. The structure and reactivity of these species are considered in popular practice to be biradical in character (5). It is proposed that these reactive intermediates are responsible for the oxidation of the sulfur dioxide in the system. The products of the initial reaction are stable aldehydes and the active biradical species according to the equations

$$O_3 + CH_3-CH=CH \rightarrow HCHO + CH_3\overset{\cdot}{C}HOO\cdot \qquad (1)$$

$$\rightarrow CH_3CHO + H\overset{\bullet}{C}HOO\bullet \tag{1a}$$

The intermediate can only interact with the olefin, molecular oxygen and the sulfur dioxide. The finding that the stoichiometry of the reaction is affected by the presence of oxygen indicates that the intermediates are not directly involved in a secondary attack on the olefin. Interactions of the intermediate with molecular oxygen occur as follows:

$$R\overset{\bullet}{C}HOO\bullet + O_2 \rightarrow \bullet OH + R\overset{O}{\overset{\|}{C}}OO\bullet \tag{2}$$

$$R\overset{\bullet}{C}HOO\circ + O_2 \rightarrow O_3 + RCHO \tag{3}$$

A secondary attack on the olefin by species other than ozone generated in the system has been demonstrated and, through the calculated addition of a prescribed amount of carbon monoxide, the agent responsible for the secondary attack was tentatively identified as the hydroxyl radical. A reduction in the aldehyde yield observed in experiments conducted in a reduced oxygen atmosphere indicates that reaction 3 is also active in the system. This ozone production is apparently offset by the increasing participation of ozone in reactions involving unidentified products of the primary reaction.

An apparent decomposition reaction of the acetyl form of the biradical intermediate was noted in the propylene–ozone interactions and resulted in the production of ketene according to the reaction

$$CH_3CHOO\bullet \rightarrow CH_2=C=O + H_2O \tag{4}$$

The rate of this reaction must be slow since the ketene was not observed in systems which included sulfur dioxide. This result along with a materials balance analysis enforced the identification of the intermediate as the species which oxidized the sulfur dioxide particularly as none of the other species under consideration form ketene.

The hydroxyl radical was involved in a secondary attack on the olefin species. The rate constant for these reactions is approximately one hundred times greater than is the rate constant for the hydroxyl radical–sulfur dioxide reaction thus eliminating that species from consideration as the oxidizing agent. The propylene–hydroxyl radical reaction is itself chain propagating:

$$\bullet OH + CH_3-CH=CH_2 \rightarrow CH_3\overset{\bullet}{C}H-CH_2OH \tag{5}$$

The final step in the proposed mechanism is the oxidative consumption of the sulfur dioxide by the reactive biradical species:

$$R\overset{\bullet}{C}HOO\bullet + SO_2 \rightarrow SO_3 + RCHO \tag{6}$$

Other potentially oxidizing species for the sulfur dioxide have been eliminated as a result of the experimental observations previously discussed.

Additional mechanistic steps are advanced to account for other products observed by the infrared spectroscopic analysis.

An oxygen effect was observed in the production of formic acid thereby eliminating the decomposition of the formyl form of the intermediate as being responsible for its formation. It is speculated, however, that a reaction between the hydroxyl and the peroxyformyl radicals could produce the acid observed:

$$\bullet OH + H\overset{O}{\overset{\|}{C}}OO\bullet \rightarrow O_2 + HCOOH \tag{7}$$

The formation of carbon monoxide from another decomposition of the formyl form of the biradical species in analogy with the formation of ketene from the decomposition of the acetyl form was not observed. There was an oxygen effect noted in the formation of the carbon monoxide and, in contrast to the ketene, the presence of sulfur dioxide did not affect the observed concentration. The peroxyformyl radical requires an enriched oxygen atmosphere for its formation in this mechanism, is known to be highly unstable and on decomposition can account for the carbon monoxide as follows:

$$H\overset{O}{\overset{\|}{C}}OO\bullet \rightarrow CO + H\overset{\bullet}{O}_2 \tag{8}$$

The hydroxyl radical formed in the oxygen enriched system reacts with the aldehyde species with about the same rate constant as with the olefin. This reaction leads to the formation of formyl and acetyl radicals which in the presence of oxygen become an additional source of peroxyformyl and peroxyacetyl radicals:

$$\bullet OH + RCHO \rightarrow H_2O + R\overset{O}{\overset{\|}{C}}\bullet \tag{9}$$

$$R\overset{O}{\overset{\|}{C}}\bullet + O_2 \rightarrow R\overset{O}{\overset{\|}{C}}OO\bullet \tag{10}$$

The notable difference between the foregoing mechanism and the three-step mechanism employed by Cox and Penkett (3) is the relatively large involvement of secondary reactions shown above. The deviation from 1:1 stoichiometry results from a secondary consumption of the olefin and secondary reactions between ozone and other products of the olefin-ozone-sulfur dioxide system. The observed oxygen effect in these studies, which affected the stoichiometry and the yield of the various products, indicates that in an oxygen deprived system the consumption of ozone by product species becomes more significant relative to the olefin consumption. In an oxygen enriched system, however, additional ozone is regenerated as are the radical species.

Conclusions

1. Wall effects did not significantly affect the results or the conclusions of this study. The losses of the individual reactants were measured and found not to affect the stoichometric results. The sulfur mass balance and the aerosol reciprocal number concentration measurements indicate that a measurable loss of sulfur to the walls occurred only at the terminus of the runs when the surface to volume ratio of the reactor became excessively large. The lower recovery of total carbon evaluated near the end of the runs was believed due to the carbon content of unmeasured products.

2. The stoichiometry of the propylene–ozone reaction was found to be a smooth function of the initial concentration ratio of these species. The olefin/ozone consumption was $>$ 1 for a system in which the olefin was initially in excess and $<$1 for a system having an initial ozone excess. The consumption ratio was unaffected by the addition of sulfur dioxide to the reactor or by varying the relative humidity over the range of 20 to 38%.

3. Molecular oxygen had a significant effect on the reaction stoichiometry and product formation in the propylene–ozone thermal reaction. The propylene/ozone consumption ratio was lower in a system in which the molecular oxygen concentration was reduced. Oxygen also contributed to the regeneration of ozone and the production of the hydroxyl radical species, both of which interacted with the propylene and with products of the reaction. Although the involvement of the hydroxyl radical had been postulated, these experiments provided the first direct evidence of its participation and its role.

4. Acetyaldehyde, formaldehyde, carbon monoxide, ketene and formic acid were observed products of the propylene–ozone reaction. The concentrations of acetyaldehyde, formaldehyde and carbon monoxide were affected by the molecular oxygen concentration. Ketene was apparently unaffected by the availability of oxygen while the production of formic acid depended on its presence. Ketene was not observed in any of the reactions involving sulfur dioxide as a reactant indicating that its rate of formation is significantly slower than the rate of the bimolecular reaction between sulfur dioxide and the intermediate species.

5. A mechanism has been advanced for the oxidative consump-

tion of sulfur dioxide in the propylene-ozone-sulfur dioxide system and which is consistent both internally and with the data observed in this study. Secondary reactions are a distinctive feature of this model which incorporates relatively few reactions to explain the major characteristics of the system studied.

Literature Cited

1. Environmental Protection Agency, "Health Consequences of Sulfur Oxides: A Review from Chess, 1970-1971," Environmental Health Effects Research Series, EPA-650/1-74-004, Research Triangle Park, North Carolina, 1974.

2. Groblicki, P. J. and Nebel, G. J., "The Photochemical Formation of Aerosols in Urban Atmospheres." Chemical Reactions in Urban Atmospheres, (C.S. Tuesday, ed.), p. 241-267, American Elsevier, New York, 1971.

3. Cox, R. A. and Penkett, S. A., Journal of the Chemical Society, Faraday Transactions I, Part 9, (1972), 68, p. 1735.

4. Environmental Protection Agency, "Aerosol Formation from Gas-Phase Reactions of Ozone and Olefin in the Presence of Sulfur Dioxide," Environmental Monitoring Series, EPA-650/ 4-74-034, Research Triangle Park, North Carolina, 1974.

5. Calvert, J. G., "Interactions of Air Pollutants," Presented at the National Academy of Sciences, Conference on Health Effects of Air Pollutants, Washington, DC, 1973.

INDEX

INDEX

A

Acetate filter rods 83
Acetyaldehyde production 188, 196
Adsorbent, carbon 10, 113
Adsorption
  kinetics equation 111
  parameters for GB vapor on carbon   118
  test apparatus, vapor 113
Advection rate 22
Aerosol(s) 2
  analysis 161
  dioctylpthalate 73
  filtration
    dielectrophoresis in 68
    electrically augmented 68
    by fibrous filter mats 91
  heterogeneous removal of free
    radicals by 17
  inclusion 19
  interactions, gas– 18
  interactions, pesticide– 52
  model, water–shell 20
  propellants in submarine
    atmosphere 7
  removal term 24
  size 100
  sulfate 187
  tropospheric 18
  in urban atmospheres 18
Air
  pesticides in 47, 48
  plenum 119
  quality
    data, St. Louis, Mo. 161
    data, West Covina, Calif. 166
    diurnal 161
  trace gas adsorption from
    contaminated 110
  velocity, dielectrophoretic augmen-
    tation factor as a function of 72
Airflow conditions, GB vapor
  breakthru of carbon under 115
Aldrin 53
Alicyclics 5
Aliphatic hydrocarbons 4
Alkyl nitrates 132, 160
Ambient halocarbons 152
Ammonia 159
Anthropogenic pollution 179
Aromatic hydrocarbons in
  submarine atmospheres 5

Arsenic exposure 65
Atmosphere(s)
  nonurban 174
  pestcides in the 42, 51
  submarine 1, 4, 5, 7, 13
  urban 18, 159
Atmospheric residence times of
  pesticides 55
Augmentation factor,
  dielectrophoretic 72

B

Benzene exposure 65
Benzyl nitrite 132, 139
Biologic monitoring of
  urine or blood 65
Blood, biologic monitoring of 65
Blood lead determinations 65
Burner, catalytic 3

C

California, West Covina 166, 169
Carbon
  adsorbent, activated 113
  adsorbent bed 10
  adsorption of GB vapor on 115, 118
  balance of organic nitrite
    photolysis 149
  dioxide 2
  monoxide 2
    continuum 193
    scavenger 190
    tracer 168
  tetrachloride 153
Catalytic burner 3
Cellulose acetate continuous filaments,
  crimped 79
Cements 3
Charcoal, activated 10
Chlordane 58
Chlorinated compounds 7
Cigarette
  filter 80, 81, 87
  products from regular size 7
  smoke filtration 79
Coal-fired power boiler 106
Coefficient of correlation 119
Contaminants
  in an exhaust control line, gaseous   119
  in the occupational environment   63

Contaminants *(continued)*
  precipitation scavenging of
    organic 28
  in submarine atmospheres 1, 6, 12
Contaminated air, trace gas
    adsorption from 110
Cooking, by-products of 3
Correlation, coefficient of 119
Coshocton, Ohio 175
Criegee mechanism 196

D

Data handling 30
DDT 43
Deactivators 127
Dieldrin 46
Dielectrophoresis in aerosol filtration .. 68
Dielectrophoretic augmentation
    factor 72
Diesel fuel 6
Diethylamine 29, 36
Diffusion 92
Diffusivity coefficient 46
Diostylphthalate (DOP) 71, 73, 91, 100
Dithizone method, blood lead
    determinations by the 65
Diurnal
  air quality 161
  pattern of free radicals 19
  variation 46
Dorman parameters for fibrous
    filter mats 102

E

Efficiency equation, single fiber 84
Electrode assembly, filter 70
Electrostatic precipitators 6
Employee health hazards 63
Endrin 58
EPAEC scavenging model 29
Erosion, wind 44
Ethylacetoacetate 29, 34
Ethyl alcohol 3
Ethyl nitrite 137
Evaporation rate of a pesticide 44
Exhaust control line, gaseous
    contaminants in 119
Exhaust filter system 119
Eye irritant 132

F

Fiber, effective radius of a 84
Fiber efficiency, single 84
Fibrous filter mats 91, 96, 102
Filter(s)
  cigarette 80, 81, 87
  –electrode assembly 70

Filter(s) *(continued)*
  glass-fiber 70, 172
  mats, fibrous 91, 96, 102
  mats, physical, properties of 97
  media, glass fiber 70
  medium, retention of
    dioctylpthalate (DOP)
    aerosol by 73
  penetration–velocity profiles 91
  rods, acetate 83
  system, exhaust 119
Filtration
  aerosol 68, 91
  cigarette smoke 79
  mechanisms 83
  of the smoke fixture 89
  theory 82
Fires, submarine 14
Flatus 3
Flue gas scrubbing 107
Formaldehyde 196
Fourier transform infrared
    spectrometer 189
Free radicals 17, 19, 24
Fuels 3, 6

G

Gas
  adsorption, trace 110
  interaction, aerosol– 18
  phase
    kinetics 17
    organic nitrates 160
    products of methyl nitrite
      photolysis 149
    reactions of ozone and olefin 187
  on rate constant, effect of added 124
  washout behavior of trace 28
Gaseous contaminants in an
    exhaust control line, ... 119
GB (isopropyl methylphosphono-
    fluridate) 110, 115, 118
Generation rates of organic
    contaminants 12
Glass fiber filters 70, 172

H

Halocarbons 152, 154, 155
Halogenated compounds 12
Health hazards, employee 63
Henry's constants of pesticides 57
Heptachlor 46
n-Hexane concentration in
    submarine atmosphere 13
Hopcalite-catalyzed oxidation of
    compounds 12
Hopcalite, decomposition of
    halogenated compounds over 12

Hydrocarbon(s)
  analyzer, NRL total 10
  photooxidation of 132
  in submarine atmospheres 4, 5, 13
Hydrogen fluoride 2
Hydrogen sulfide reaction, ozone 122
Hydroxyl radical 183, 192
Hygroscopic layer 18

I

Inertial impaction 92
Infrared spectra of the
  organic nitrites 139
Infrared spectrometer, Fourier
  transform 189
Inorganics in submarine atmospheres ... 4
Insecticides, organochlorine 43
Interception 92
Isopentane 175
Isopropyl methylphosphonofluori-
  date (GB) 110, 115, 118

K

Kane, Penn. 175
Ketene spectrum 193
Kinetics
  equation, adsorption 111
  gas phase 17
  of ozone-hydrogen sulfide
    reaction 122
  of trace gas adsorption from
    contaminated air 110

L

Lead determinations, blood 65
Lead exposure 65
Levulinic acid dehydratase (ALAD)
  activities 65
Lewisburg, W. Va. 175
Lindane 55
Linear face velocity range 92

M

Magnesium oxide process 107
Magnitude effects of aerosol inclusion ... 19
McHenry, Md. 175
Meteorological conditions affecting
  pesticide transport 50
Meteorological sensors 30
Methane 3
Methyl chloroform 10
Methyl iodide 153
Methyl nitrite 135, 149
Missouri, St. Louis 162, 169
Molecular oxygen concentration 190

Monoethanolamine (MEA)
  scrubber 2

N

National Institute for Occupational
  Safety and Health Act 63
New Jersey metropolitan region 18
New York City 157
New York metropolitan region 18
Nicotine 80, 85, 86
Nitrates, gas phase organic 160
Nitric acid 136, 159
Nitric oxide 19, 64
Nitrite photolysis, organic 146, 149
Nitrites, organic 139
Nitrogen
  compounds, Hopcalite-catalyzed
    oxidation of 12
  dioxide 2, 64, 175
  oxides 159, 169, 170, 175
  pressurization with 14
Nitrous oxide 19
Nonurban atmospheres 174
NRL total hydrocarbon analyzer 10
Nuclear submarines, atmosphere of .. 1

O

Occupational environment, hazardous
  contaminants in the 63
Occupational Safety and Health Act .. 63
Ohio, Coshocton 175
Oils 3
Olefin–ozone reactions 187
Olympic Peninsula of western Wash. .. 29
Organics
  compounds, large-molecule 10
  contaminants, precipitation
    scavenging of 28
  contaminants in submarine, ... 5, 12
  nitrates, gas phase 160
  nitrite photolysis 146, 149
  nitrites, infrared spectra of 139
  vapor removal system 11
Organochlorine insecticides 43
Oxidation, free radical 24
Oxygen concentration, molecular 190
Ozone 2, 22, 159
  generation, ... 184
  hydrogen sulfide reaction 122
  measurements, vertical 178
  monitoring stations 177
  nighttime concentrations of 183
  in nonurban atmospheres 174
  –olefin reactions 187
  –propylene reaction 192
  –sulfur dioxide system, propylene– ... 196
  synthesis 175
Ozonide ring, five-membered 129

P

Paints 3, 6
Parathion 58
Particle removal efficiency 82, 86
Particle sizes of pesticides 49
Pennsylvania, Kane 175
Perchlorethylene 153
Peroxyacetyl compound 137
Peroxyacetyl nitrate 132
Peroxyacyl nitrate formation 139
Peroxybenzoyl nitrate 132
Peroxyformyl nitrate 133
Peroxyformyl radical 198
Peroxypropionyl nitrate 132
Personal protective equipment 65
Pesticide(s)
  –aerosol interactions 52
  in air, sources of 48
  in the atmosphere, persistent 42
  atmospheric residence times and
     transport potentials of 55
  dispersion in the air 47
  dry deposition of 54
  evaporation rate of 44
  flux 46
  Henry's constants of 57
  losses during application 44
  particle sizes of 49
  photochemical degradation of 53
  precipitation scavenging of 55
  removal from the atmosphere 51
  research needs 58
  transport 44, 48, 50
Phenol 80
Phosgene 157
Photochemical
  degradation of pesticides 53
  oxidant 174
  reactions, chamber for 134
  smog 172, 187
  systems, "spent" 183
Photolysis
  of alkyl nitrites and benzyl nitrite 132
  methyl nitrite 149
  organic nitrite 146, 149
Photooxidation of hydrocarbons 132
Plenum, air 119
Pollution, anthropogenic 179
Polychlorinated biphenyl 54
Power boiler, coal-fired 106
Precipitation scavenging 28, 55
Precipitators, electrostatic 6
Pressure drop of acetate filter rods 83
Pressure drop equation 80, 81
Pressurization with nitrogen 14
Product ratios 125
Propellants, aerosol 7
Propylene
  loss 188

Propylene (continued)
  –ozone reaction 192
  –ozone–sulfur dioxide system 196
Propyl nitrite 137
Protective equipment, personal 65
Pyrolysis products form otbacco 2

R

Radical(s)
  free 17, 19, 24
  hydroxyl 183, 192
  impact removal of 20
  peroxyformyl 198
  removal rate for 19
Rain samplers 30
Raindrop spectra 30
Rainfall washout 56
Rainwater
  concentrations 35
  diethylamine in 36
  ethylacetoacetate in 34
Rate constants
  1.5 order 124
  second order 128
  3/2 order 127
Rate law 126
Refrigerants in submarine
     atmospheres 7
Refrigeration systems 3
Research needs, pesticide 58
Residence times of pesticides,
     atmospheric 55
Respirators 65

S

Scavenger, carbon monoxide 190
Scavenging model, EPAEC 29
Scavenging, precipitation 28, 55
Scrubber, monoethanolamine (MEA) ... 2
Sealers 3
Sensors, meteorological 30
Ship's machinery, hot oils in 3
Smog
  chamber experiments 153, 176
  photochemical 172, 187
  urban 132
Smoke 6
  filtration, cigarette 79
  mixture, selective filtration of the .. 89
  removal efficiency 85
Smoking, by-products of 3
Smoking machine, standard 7
Solubility, tracer 32
Solvents 3
Spectrometer, Fourier transform
     infrared 189
Spent photochemical system 183

Stabilization effects of aerosol
  inclusion 19
Standards development in the control
  of hazardous contaminants in
  the occupational environment 63
Standard smoking machine 7
St. Louis, Mo. 161, 169
Stratospheric reactivity 157
Submarine(s)
  atmospheres, nuclear 1
  aliphatic hydrocarbons in 4
  aromatic hydrocarbons in 5
  chlorinated compounds in 7
  n-hexane concentration as a
    function of total hydro-
    carbon adsorbed in 13
  inorganics in 4
  organics in 5, 12
  refrigerants and aerosol
    propellants in 7
  removal of contaminants from 1
  unsaturates and alicyclics in 5
  contamination control in 6
  fires 14
Sulfate aerosols 187
Sulfur dioxide
  gas-phase ozone–olefin reactions
    in the presence of 187
  gas from sulfuric acid mist,
    separation of 66
  removal 106
  system, propylene–ozone– 196
  washout 28
Sulfur recovery 106
Sulfuric acid mist, separation of
  sulfur dioxide gas from 66
System behavior, proposed 181

T

Tar removal efficiency 85
Time dependence of the nitrogen
  oxide loss profile 170
Tobacco, pyrolysis products from 2
Toluene exposure 65
Trace gas adsorption from
  contaminated air, ... 110
Trace gases and vapors, washout
  behavior of 28
Tracer
  analysis 30
  concentration 38

Tracer *(continued)*
  release equipment 30
  solubility 30
Transition state 129
Transport, meteorological conditions
  affecting 50
Transport potentials of pesticides 55
Trichlorethylene 10
Trichloroethane 153
Trichlorofluoromethane 153
Trichloromethane 153
Troposphere, urban 17
Tropospheric aerosol 18
Tropospheric reactivity 157

U

Unsaturates in submarine
  atmospheres 5
Urban
  atmospheres, aerosols in 18
  atmospheres, nitrogen oxides in 159
  smog 132
  troposphere 17
Urine, biologic monitoring of 65

V

Vapor(s)
  adsorption test apparatus 113
  on carbon adsorption of GB 115, 118
  dispersion equipment 31
  removal system, organic 11
  samplers 30
  washout behavior of 28
Velocity profiles, filter 92
Velocity range, linear face 92
Vinylidene chloride 10
Voltage, dielectrophoretic augmen-
  tation factor as a function of 72

W

Washington, Olympic Peninsula
  of western 29
Washout
  behavior of trace gases and vapors .. 28
  rainfall 56
  sulfur dioxide 28
Water–shell aerosol model for impact
  removal of radicals 20
West Covina, Calif. 166, 169
West Virginia, Lewisburg 175
Wind erosion, transport of
  pesticides by 44

Removal of Trace Contaminants from the Air

A symposium co-sponsored by the Division of Colloid and Surface Chemistry and the Division of Environmental Chemistry of the American Chemical Society

Controlling air pollution is a complex problem since each pollutant must be identified and its basic chemistry and physics understood before a method can be designed to remove it from the atmosphere. Added to this is the fact that there are innumerable kinds of air pollution depending on the particular environment in question.

This volume discusses air pollution characterization in and removal from many various situations. Specifically, 16 chapters cover contaminants in submarine atmospheres, removal of free radicals by aerosols, precipitation scavenging of organic contaminants, persistent pesticides in the atmosphere, hazardous contaminants in the occupational environment, electrically augmented filtration of aerosols, cigarette smoke filtration, aerosol filtration by fibrous filter mats, sulfur dioxide in stack gases, trace gas adsorption from contaminated air, ozone–hydrogen sulfide reactions, photolysis of alkyl nitrites and benzyl nitrite, ambient halocarbons, nitrogen oxides in urban atmospheres, ozone concentrations in nonurban atmospheres, and ozone–olefin reactions.